T0234815

Solid-State Electrochemistry

Abdelkader Hammou · Samuel Georges

Solid-State Electrochemistry

Essential Course Notes and Solved Exercises

 UGA Editions

 Springer

Abdelkader Hammou
Laboratoire d'Electrochimie et de
Physico-chimie des Matériaux et des
Interfaces
PHELMA
Saint-Martin-d'Hères, France

Samuel Georges
Laboratoire d'Electrochimie et de
Physico-chimie des Matériaux et des
Interfaces
PHELMA
Saint-Martin-d'Hères, France

This translation has been supported by **UGA Éditions**, publishing house of Université Grenoble Alpes and the **région Auvergne-Rhône-Alpes**. https://www.uga-editions.com/.

ISBN 978-3-030-39661-9   ISBN 978-3-030-39659-6   (eBook)
https://doi.org/10.1007/978-3-030-39659-6

Translated, revised and adapted from "Electrochimie des solides", A. Hammou/S. Georges, EDP Sciences, 2011.

This Springer imprint is published by the registered company Springer Nature Switzerland AG
The registered company address is: Gewerbestrasse 11, 6330 Cham, Switzerland

Grenoble Sciences

The French version of this book received the "Grenoble Sciences" label. "Grenoble Sciences" directed by Professor Jean Bornarel, was between 1990 and 2017 an expertising and labelling centre for scientific works, with a national accreditation in France. Its purpose was to select the most original high standard projects with the help of anonymous referees, then submit them to reading comittees that interacted with the authors to improve the quality of the manuscripts as long as necessary. Finally, an adequate scientific publisher was entrusted to publish the selected works worldwide.

About this Book

This book is translated, revised and adapted from *Électrochimie des solides – Exercices corrigés avec rappels de cours* by Abdelkader Hammou & Samuel Georges, EDP Sciences, Grenoble Sciences Series, 2011, ISBN 978-2-7598-0658-4.

The Translation from original French version has been performed by: Brett Kraabel, Physical Sciences Communication.

The reading committee of the French version included the following members:

▷ J.M. Bassat, CNRS research director, Bordeaux

▷ J. Fouletier, professor at Grenoble Alpes University

▷ R.N. Vannier, professor at the ENS of Chemistry, Lille

and K. Girona, PhD student

Preface

The electrochemistry of solids is a relatively young field that only really caught on in the 1950s. Its particularity resides in the multidisciplinary nature of its content, which associates electrochemistry, solid-state chemistry (organic and inorganic), and physical chemistry. The principal goal is to synthesize and characterize materials susceptible for use in devices that exploit their electro-chemical properties. The materials studied are either electrolytes or electrode materials, which are two major families of materials.

▷ The electrolytes discussed are solid oxide or halide solutions in the crystalline state [$(ZrO_2)_{1-x}(Y_2O_3)_x$, β alumina, $(SrCl_2)_{1-x}(KCl)_x$, ...], or glassy state (SiO_2-K_2O, LiI, $Li_4P_2S_7$, ...) and organic-polymer-salt complexes (POE-LiTFSI). In this work, we mainly study the structural characteristics (i.e., phases, crystalline structure) and the ionic transport properties (i.e., electrical conductivity, transport modes, ionic domain). The experimental results are analyzed by considering the properties under study as a function of temperature, of the nature and concentration of structural defects (vacancies, interstitials, impurities, dopants) present in the phase, and of the chemical potential of the basic constituents of the phase. An example of the latter is oxygen in the case of solid oxide electrolytes; in this case, Brouwer diagrams are frequently used. Finally, note that, in most solid electrolytes, the conductivity derives only from one ionic species.

▷ Electrode materials are made of metals (Li, Na, Ag, Pt, ...) as well as oxides ($La_{1-x}Sr_xMnO_{3-\delta}$, $FePO_4$, WO_3, ...), composites (Ni-YSZ), or sulfides (TiS_2, MoS_2, ...). Research in this field focuses primarily on determining the electrical conductivity and identifying and studying the kinetics of the electrode reactions and the durability of the electrode-electrolyte interface. As is the case with aqueous solutions, the results are interpreted by relying on the polarization due to adsorption-desorption, to diffusion and migration of electroactive species, and to charge transfer.

Today, the international electrochemistry community is well structured, with regular conferences and a large volume of significant publications in electrochemical journals (*Journal of the Electrochemical Society, Journal of Power Sources, Solid State Ionics, Ionics*) and in solid-state chemistry journals (*Journal of Materials Science, Journal of the European Ceramic Society, Journal of the American Ceramic Society, …*). Teaching texts in this field, however, are few and take the form of courses, of chapters written by specialists, or of proceedings of conferences dealing with the solid-state electrochemistry.

The origin of this collection of exercises is the desire to provide a work tool in the form of exercises to satisfy the need expressed by doctoral students in our laboratory and by participants in the continuing-education courses on solid-state electrochemistry organized by our group at the Laboratoire d'Électrochimie et de Physicochimie des Matériaux et des Interfaces of Grenoble (LEPMI). To the best of our knowledge, no such work exists to this day. Our goal is thus to fill a void by allowing readers to familiarize themselves, by solving problems, with the notions presented in Solid-State Electrochemistry. These problems cover essentially

- ▷ the notation of defects in ionic crystalline solids with the focus on the notion of effective charge,
- ▷ the evolution of the stoichiometry as a function of temperature, of the doping level, and of the chemical potential of the basic constituents of the materials under study, in particular by using Brouwer diagrams,
- ▷ methods to measure electrochemical quantities (conductivity, transport number, electrode polarization) such as impedance measurements, dilatocoulometry, and drawing current-voltage curves,
- ▷ the study of several applications involving solid electrolytes, such as fuel cells, batteries, and sensors.

The essential skills required to solve these exercises are presented in the form of course notes. These are given at the outset of each chapter. To delve deeper into a question, one should consult the specialized books and articles in the (non-exhaustive) bibliography that appears at the end of this book. The publications from which some of the exercises in this book were constructed also appear in the bibliography.

We hope that our contribution will illuminate with a less arduous light this field towards which we hope to attract a larger public.

We are grateful to Professor Jacques Fouletier and Professor Pierre Fabry for having provided us with a certain number of the exercises proposed herein; exercises that were used in the exam for the Master II diploma in *Electrochemistry and Materials* from the Université Grenoble Alpes. We also thank Elisabeth Siebert, Cécile Rossignol, and Jean-Louis Souquet for reading the original manuscript, discussing the pertinence of the exercises, and especially for taking time out from their schedule to verify the answers. Finally, our gratitude goes out to all our colleagues who, after reading the manuscript, gave us their opinion and constructive recommendations; in particular Rose-Noëlle Vannier, Jean-Marc Bassat, and Jacques Fouletier, as well as Kelly Girona.

Table of contents

Base quantities, units, and symbols from the international system (IS)

Quantity	Symbol	Units	Symbol
Absolute temperature	T	kelvin	K
Amount of substance	n	mole	mol
Electric current	I	ampere	A
Length	ℓ	meter	m
Luminous intensity	Iv	candela	cd
Mass	m	kilogram	kg
Time	t	second	s

Physical and chemical constants

Avogadro constant	$N_{Av} = 6.02 \times 10^{23}\ mol^{-1}$
Boltzmann constant	$k = 1.38 \times 10^{-23}\ J\ K^{-1}$
Electron charge	$e = 1.6 \times 10^{-19}\ C$
Faraday constant	$F = 96\,480\ C\ mol^{-1}$
Ideal gas constant	$R = 8.314\ J\ mol^{-1}\ K^{-1}$
Imaginary unit	$j^2 = -1$
Pi	$\pi = 3.14159$
Speed of light	$c = 3 \times 10^8\ m\ s^{-1}$
Standard pressure	$P^0 = 10^5\ Pa$
Vacuum permittivity	$\varepsilon_0 = 8.85 \times 10^{-12}\ F\ m^{-1}$

Conversions

Energy	$1\ eV = 1.6 \times 10^{-19}\ J$
Pressure	$1\ bar = 10^5\ Pa;\ 1\ atm = 1.013 \times 10^5\ Pa$
Temperature	$T\ [K] = T\ [^\circ C] + 273$

Physical-chemistry Symbols and units

Physical-chemistry quantity	Symbol	Units
Activation energy	E_a	$J\,mol^{-1}$, eV
Activity of species i	a_i	–
Admittance	Y	S
Angular velocity	ω	$rad\,s^{-1}$
Capacity	C	F
Chemical potential	μ	$J\,mol^{-1}$
Concentration of species i	[i], C_i	$mol\,m^{-3}$, $mol\,L^{-1}$
Conductance	G	S
Current density, (exchange current)	i, (i_0)	$A\,m^{-2}$, $A\,cm^{-2}$
Diffusion coefficient	D	$cm^2\,s^{-1}$
Electric charge	Q	C
Electric mobility	u	$cm^2\,V^{-1}\,s^{-1}$
Electric resistance	R	Ω
Electrical conductivity	σ	$S\,cm^{-1}$
Electrochemical mobility	$\tilde{u}$	$cm^2\,s^{-1}\,J^{-1}\,mol$
Electrochemical potential	$\tilde{\mu}$	$J\,mol^{-1}$
Electromotive force (emf)	ΔE	V
Electrostatic potential	φ	V
Enthalpy of formation	$\Delta_f H$	$J\,mol^{-1}$
Entropy of formation	$\Delta_f S$	$J\,mol^{-1}\,K^{-1}$
Equilibrium constant	K	–

Physical-chemistry quantity	Symbol	Units
Flow	D	$L\,s^{-1}$
Flux density	J	$mol\,s^{-1}\,m^{-2}$
Frequency	f, ν	$Hz,\,s^{-1}$
Geometric factor	k	$m^{-1},\,cm^{-1}$
Gibbs free energy of formation	$\Delta_f G$	$J\,mol^{-1}$
Impedance	Z	Ω
Inductance	L	H
Lattice parameter	a, b, c, α, β, γ	Å
Magnitude of X in standard state	X°	–
Molar heat capacity at constant pressure	C_p	$J\,K^{-1}\,mol^{-1}$
Molar mass	M	$g\,mol^{-1}$
Molar volume	V_m	$m^3\,mol^{-1}$
Mole fraction	x	–
Overpotential	η	V
Permittivity	ε	$F\,m^{-1}$
Potential difference	ΔV, ΔE	V
Power	P	W
Power density	P	$W\,cm^{-2}$
Pressure	P	Pa, bar
Relative permittivity	ε_r	–
Fractional coverage	θ	–
Resistivity	ρ	$\Omega\,cm$
Specific energy	W	$J\,kg^{-1},\,W\,h\,kg^{-1}$
Thermal expansion coefficient	α	K^{-1}
Transport number of species i	t_i	–

Acronyms and abbreviations used in this book

AE	Auxiliary electrode
AFC	Alkaline fuel cell
BICUVOX	Bismuth copper vanadium oxide
CE	Counter electrode
Cermet	Ceramic metal composite
CGP	Praseodymia and gadolinia doped ceria
CIS	Complex impedance spectroscopy
const.	Constant
CPE	Constant phase element
DIR	Direct internal reforming (of a hydrocarbon)
emf	Electromotive force
expt	Experimental
fcc	Face-centered cubic
GDC	Gadolinia-doped ceria
GIR	Gradual internal reforming (of a hydrocarbon)
log	base 10 logarithm
LSCo	Lanthanum strontium cobaltite
LSGM	Lanthanum gallate doped with strontium and metal oxides
LSM	Lanthanum strontium manganite
M	Molar mass
MIEC	Mixed ionic-electronic conductor
NAFION	Proton-conducting polymer used as electrolyte in proton-exchange membrane fuel cells

NASICON	Solid solution rendered conductive by Na^+ ions; chemical formula is $NaZr_2Si_2PO_{12}$
Ni-YSZ	Cermet composed of nickel-yttria stabilized zirconia
NTP	Normal temperature and pressure
Oh	Octahedral (environment)
PEMFC	Proton-exchange membrane fuel cell
PEO	Polyethylene oxide
R//C	parallel RC circuit
RE	Reference electrode
SE	Solid electrolyte
SIMS	Secondary ion mass spectrometry
SOFC	Solid oxide fuel cell
STP	Standard temperature and pressure.
Td	Tetrahedral (environment)
TFSI	Trifluoromethanesulfonyl-imide
V_m	Molar volume
VTF	Vogel-Tammann-Fulcher (model)
W	Warburg (impedance, element)
WE	Working electrode
YSZ	Yttria stabilized zirconia
ΔV	Potential difference

Chapter 1

Description of ionic crystals

Course notes

1.1 – Definitions

1.1.1 – The perfect crystal

The notion of the perfect crystal is based on the results of crystallography. This discipline describes a solid by giving, in particular, the precise location of points in space (called lattice sites) occupied by the chemical species of the crystal. To illustrate this notion, consider the example of an ionic crystal described by the formula MX (M^+, X^-). The perfect MX crystal consists of a lattice of normal sites that are occupied, between which are unoccupied interstitial sites. We distinguish between the normal cationic sites, all occupied by the species M^+, and the normal anionic sites, all occupied by the species X^-. Figure 1(a) shows a schematic two-dimensional representation of a perfect MX crystal. Note that this pure stoichiometric composition does not exist, although we can approach it near 0 K with extremely pure material.

1.1.2 – The real crystal

For a real crystal, we consider thermal motion and the presence of foreign elements. Thermal motion is responsible for the formation of defects, which results from the displacement of the original chemical elements [see fig. 1(b)]. For a pure real crystal, we normally identify M^+ and X^- vacancies, which are denoted V_M and V_X, respectively, interstitial species, denoted M_i^+ and X_i^-, and antistructure defects M_X^+ and X_M^-, which are extremely unlikely to form

© Springer Nature Switzerland AG 2020
A. Hammou and S. Georges, *Solid-State Electrochemistry*,
https://doi.org/10.1007/978-3-030-39659-6_1

in ionic crystals. Such defects are referred to as "intrinsic." Foreign elements, which consist of impurities or dopants, occupy either normal sites or interstitial sites with respect to the perfect crystal. For example, the presence of the impurity DY (D^{2+}, Y^{2-}) in MX leads to D_M^{2+} and Y_X^{2-} defects by substitution and to D_i^{2+} and Y_i^{2-} defects by insertion. Defects related to the presence of foreign elements are qualified as "extrinsic."

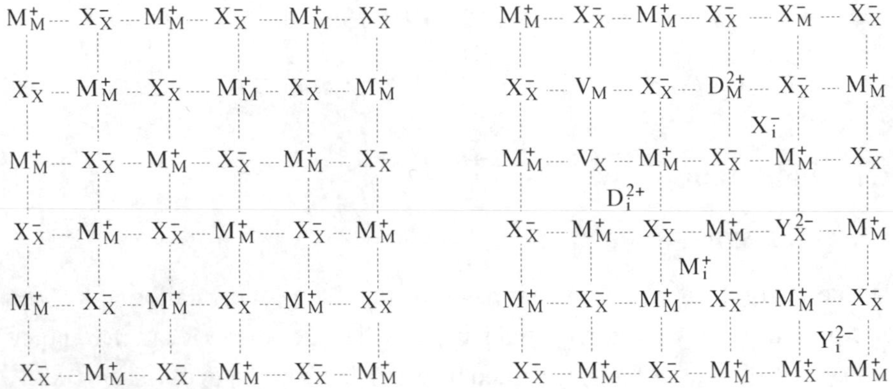

$$
\begin{array}{llllll}
M_M^+ & X_X^- & M_M^+ & X_X^- & M_M^+ & X_X^- \\
X_X^- & M_M^+ & X_X^- & M_M^+ & X_X^- & M_M^+ \\
M_M^+ & X_X^- & M_M^+ & X_X^- & M_M^+ & X_X^- \\
X_X^- & M_M^+ & X_X^- & M_M^+ & X_X^- & M_M^+ \\
M_M^+ & X_X^- & M_M^+ & X_X^- & M_M^+ & X_X^- \\
X_X^- & M_M^+ & X_X^- & M_M^+ & X_X^- & M_M^+
\end{array}
$$

$$
\begin{array}{lllllll}
M_M^+ & X_X^- & M_M^+ & X_X^- & X_M^- & X_X^- \\
X_X^- & V_M & X_X^- & D_M^{2+} & X_X^- & M_M^+ \\
 & & & & X_i^- & \\
M_M^+ & V_X & M_M^+ & X_X^- & M_M^+ & X_X^- \\
 & D_i^{2+} & & & & \\
X_X^- & M_M^+ & X_X^- & M_M^+ & Y_X^{2-} & M_M^+ \\
 & & & M_i^+ & & \\
M_M^+ & X_X^- & M_M^+ & X_X^- & M_M^+ & X_X^- \\
 & & & & & Y_i^{2-} \\
X_X^- & M_M^+ & X_X^- & M_M^+ & M_X^+ & M_M^+
\end{array}
$$

(a) **Perfect crystal**   (b) **Real crystal**

Figure 1 – Two-dimensional schematic representation of
(a) a perfect MX crystal (M^+, X^-) and
(b) a real MX crystal (M^+, X^-) that contains impurity DY (D^{2+}, Y^{2-}).

1.1.3 – Structure elements and effective charge

In chemistry and in solid-state electrochemistry, we often use a notation involving structure elements to express reactions. In this book, we use the Kröger-Vink notation, which uses structure elements. A structure element reveals a crystallographic site, the chemical species that occupies the site (or its absence), and the effective charge Q_e:

$$\text{Species}_{\text{site}}^{\text{effective charge}}$$

The effective charge is given by

$$Q_e = Q_r - Q_n$$

where Q_r is the charge of the species that actually occupies the site (i.e., the "real" charge) and Q_n is the charge of the species that would occupy the site in a perfect crystal (i.e., the "normal" charge). The effective charge may be positive, negative, or neutral. The symbols used for the sign of the charge are given in table 1.

Table 1 – Symbols used to indicate sign of effective charge.

Sign	> 0	$= 0$	< 0
Symbol	•	×	'

For example, let us write the structure element for a cation vacancy M^+ in MX. The corresponding site is a M site, the effective charge is $Q_e = 0 - 1 = -1$, and the structure element is V'_M. The structure element for the impurity D^{2+} that is substituted for M^+ in MX is denoted $D^{\bullet}_M$.

We find

▷ normal structure elements that describe the perfect crystal (i.e., $M^{\times}_M$ and $X^{\times}_X$) and

▷ structure defects (or point defects) that encompass all the other structure elements in the real crystal and that coexist with the normal structure elements.

1.2 – Reactions and equilibria

1.2.1 – Atomic disorder and electronic disorder

For thermodynamic reasons, atomic-defect pairs, which are qualified as intrinsic disorder, are the dominant type of disorder in ionic crystals. We typically identify

▷ Schottky disorder, which is characterized by the formation of cationic and anionic vacancies. In MX, this is the pair $V'_M + V^{\bullet}_X$.

▷ Frenkel disorder, which is
  • Cationic when it involves the formation of cationic vacancies and interstitial cations. In MX, an example of this is the pair $V'_M + M^{\bullet}_i$.
  • Anionic disorder when anionic vacancies and interstitial anions appear. In MX, an example of this is the pair $V^{\bullet}_X + X'_i$.

Each of these types of disorder gives rise to a chemical equilibrium involving the defects with an equilibrium constant K.

Moreover, ionic crystals always contain intrinsic electronic disorder, which involves the equilibrium in the crystal between free electrons e' and holes $h^{\bullet}$.

1.2.2 – Writing the reactions

In the electrochemistry of crystalline solids, a written reaction must satisfy:

▷ mass balance;

▷ electroneutrality, which involves the effective charges of the species that partake in the reaction;

▷ sitoneutrality: the ratio of the number of cationic sites to the number of anionic sites is conserved, with the perfect crystal serving as reference. For example, for a crystal with the formula MX_2 (M^{2+}, $2X^-$), all written reactions must satisfy the condition

$$\frac{\text{Number of cationic sites}}{\text{Number of anionic sites}} = \frac{1}{2}$$

The application to the intrinsic disorder introduced above gives the following reactions:

▷ for Frenkel cationic disorder

$$M_M^\times \rightleftharpoons V_M' + M_i^\bullet \qquad \text{with } K_{CF} = [V_M'] [M_i^\bullet]$$

▷ for Frenkel anionic disorder

$$X_X^\times \rightleftharpoons V_X^\bullet + X_i' \qquad \text{with } K_{AF} = [V_X^\bullet] [X_i']$$

▷ for Schottky disorder

$$0 \rightleftharpoons V_M' + V_X^\bullet \qquad \text{with } K_S = [V_M'] [V_X^\bullet]$$

where K_{CF}, K_{AF}, and K_S are the respective equilibrium constants.

The activity of the normal structure elements are taken to be unity. A real crystal is considered to be a dilute solution of point defects, with the solvent being the perfect crystal.

For electronic disorder, we write

$$0 \rightleftharpoons e' + h^\bullet \qquad \text{with } K_e = [e'] [h^\bullet]$$

By using $[e'] = n$ and $[h^\bullet] = p$, we obtain $K_e = np$.

1.2.3 – Presence of foreign atoms

The presence of foreign elements corresponds to an introduction reaction that is frequently qualified as a doping reaction (by insertion or substitution). The notation for the reaction highlights the role of the defects corresponding to the dominant atomic disorder in the crystal. For example, consider the reaction in which MX (M^+, X^-) is doped with DX_2 (D^{2+}, $2X^-$), given that Schottky disorder dominates in MX and D substitutes for M. The reaction is written as

$$DX_2 \longrightarrow D_M^{\bullet} + V_M' + 2X_X^{\times}$$

or $$DX_2 + V_X^{\bullet} \longrightarrow D_M^{\bullet} + 2X_X^{\times}$$

Note that the introduction of DX_2 increases the number of V_M' vacancies, which are said to be of extrinsic origin.

1.2.4 – Equilibrium with the environment

The influence of the environment results in an equilibrium reaction between the crystal and the environment that involves the activity of one of the chemical elements of the crystal that is also present in the environment. For example, consider a MX crystal in equilibrium with a gaseous atmosphere where the partial pressure P_{X_2} of species X_2 is fixed. The reaction is expressed as

$$X_X^{\times} \rightleftharpoons \tfrac{1}{2}X_2 + e' + V_X^{\bullet} \qquad \text{with } K_g = n\,[V_X^{\bullet}]\,P_{X_2}^{\frac{1}{2}}$$

where K_g is the equilibrium constant with the gaseous phase.

1.3 – Brouwer diagram

The Brouwer diagram is the curve expressing the variation in defect concentration of an ionic crystal as a function of the activity of one of the crystal constituents. The approach is analogous to the curve in a diagram showing the predominant dissolved species in an aqueous solution as a function of pH. Such a plot reveals in particular the ionic, electronic, or mixed characteristic of conduction in an ionic crystal. To plot the curve, one must first write the equilibrium expression related to the intrinsic disorder and to the environment.

1.3.1 – Equilibria

As an example, we treat the case of an ionic crystal MX_2, where anionic Frenkel disorder predominates, in equilibrium with the gaseous phase of X_2 at partial pressure P_{X_2}.

▷ atomic disorder

$$X_X^{\times} \rightleftharpoons V_X^{\bullet} + X_i' \qquad \text{with } K_{AF} = [V_X^{\bullet}]\,[X_i']$$

▷ electronic disorder

$$0 \rightleftharpoons e' + h^\bullet \qquad\qquad \text{with } K_e = np$$

▷ equilibrium with the environment

$$X_X^\times \rightleftharpoons \tfrac{1}{2}X_2 + e' + V_X^\bullet \quad \text{with } K_g = n[V_X^\bullet]\,P_{X_2}^{\frac{1}{2}}$$

1.3.2 – Electroneutrality relation and the Brouwer approximation

The electroneutrality relation is

$$n + [X_i'] = p + [V_X^\bullet]$$

We obtain a system of four independent equations with four unknowns: $[V_X^\bullet]$, $[X_i']$, n, and p. These unknowns may be expressed as a function of the equilibrium constants K_{AF}, K_e, K_g, and the partial pressure P_{X_2}. To solve this system, we use the Brouwer approximation, which consists of dividing the partial-pressure domain into three domains so that, in each domain, the concentrations of certain species dominate. For the electroneutrality relation, this equates to retaining a single term per domain, which gives

▷ $n = [V_X^\bullet]$ in the domain with low partial pressure of X_2,

▷ $[X_i'] = [V_X^\bullet]$ in the intermediate domain, and

▷ $[X_i'] = p$ in the domain with high partial pressure of X_2.

As an example, we develop the case of the domain of low partial pressure where electrons and vacancies dominate. Under these conditions, we have $n = [V_X^\bullet]$, which gives $K_g = n^2 P_{X_2}^{\frac{1}{2}}$. From this we deduce the concentration of structure defects and electronic defects:

$$n = K_g^{\frac{1}{2}}P_{X_2}^{-\frac{1}{4}}$$

$$[V_X^\bullet] = K_g^{\frac{1}{2}}P_{X_2}^{-\frac{1}{4}}$$

$$[X_i'] = K_{AF}\,K_g^{-\frac{1}{2}}P_{X_2}^{\frac{1}{4}}$$

$$p = K_e\,K_g^{-\frac{1}{2}}P_{X_2}^{\frac{1}{4}}$$

Similar calculations allow us to express the concentrations of various defects in the other domains.

1.3.3 – Diagram for MX_2 crystal

The diagram is obtained by plotting in logarithmic coordinates the variation in concentration of each species as a function of partial pressure P_{X_2} and at a given temperature. We can distinguish two cases: $K_{AF} \gg K_e$ and $K_e \gg K_{AF}$. Figure 2 shows the form of the Brouwer diagram for the compound MX_2 for $K_{AF} \gg K_e$. Note that, in the intermediate domain, a plateau appears coinciding with the dominance of atomic structure defects. This domain must not be confused with the domain of redox stability or with the domain of crystal ionicity (see fig. 41, page 98).

Elsewhere, the concentrations of all the species vary with P_{X_2}. Point S is defined in section 1.4.

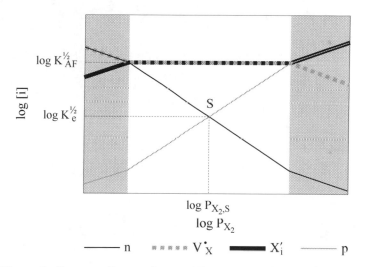

Figure 2 – Brouwer diagram for pure MX_2 compound at temperature T.

1.3.4 – Case of solid solution $(MX_2)_{1-x}(DX)_x$

We now treat the case where D is substituted for M. The reaction that substitutes DX for MX_2 is

$$DX \longrightarrow D'_M + X^\times_X + V^{\bullet}_X$$

We observe an increase in the rate of $V^{\bullet}_X$ vacancies of extrinsic origin. For typical doping levels (several %), the rate of extrinsic vacancies dominates that of the intrinsic vacancies. Under these conditions, the electroneutrality relation is

$$n + [X'_i] + [D'_M] = p + [V^{\bullet}_X] \qquad \text{where } [D'_M] = \text{const.}$$

It is interesting to consider a simplified version of this relation: $[D'_M] = [V^{\bullet}_X]$. The doping causes an increase in the size of the domain where ionic defects dominate and, consequently, of the electrolytic domain of the crystal.

1.4 – Stoichiometry and departure from stoichiometry

A crystal described by the formula M_aX_b is stoichiometric if the following condition is satisfied:

$$\frac{\text{Number of M atoms}}{\text{Number of X atoms}} = \frac{a}{b}$$

Otherwise, the compound is said to be non-stoichiometric or to depart by δ from stoichiometry (see glossary). This departure is due to the physical-chemistry conditions imposed on the crystal (temperature, P_{X_2}, ...). For example, for the compound MX_2 shown in figure 2, the stoichiometry is represented by the point S , which is characterized by the partial pressure $P_{X_2,S}$ and the temperature $T_S = T$ to which the diagram corresponds.

Exercises

Exercise 1.1 – Notation for structure elements and structure defects

1. Give the expression that relates the effective charge Q_e of a structure element or structure defect to the real charge Q_r of the ionic species occupying a site and to the charge Q_n of the species that normally occupies the site.

2. Use Kröger-Vink notation to express:
   a. a sodium vacancy in NaCl,
   b. a fluorine anion in its normal position in CaF_2,
   c. a bromine vacancy in NaBr,
   d. a magnesium cation in its normal position in MgO,
   e. an interstitial silver cation in AgBr,
   f. an oxygen vacancy in TiO_2,
   g. an oxygen vacancy that has trapped an electron in ZrO_2,
   h. an interstitial oxide ion that has trapped two holes.
   Give the values of Q_e, Q_r, and Q_n.

3. Use Kröger-Vink notation to express the structure defect that corresponds to:
   a. a Na^+ cation substituted for Mg^{2+} in MgO,
   b. an oxide ion O^{2-} substituted for F^- in SrF_2,
   c. a Al^{3+} ion in an interstitial position in CaO,
   d. a Gd^{3+} ion substituted for Ce^{4+} in CeO_2.
   Give the values for Q_e, Q_r, and Q_n.

4. Use Kröger-Vink notation to express the following associated defect pairs:
   a. a Gd^{3+} ion in a substitution position in CeO_2 and an oxygen vacancy,
   b. a chlorine vacancy and a sodium vacancy in NaCl.
   Give the values of Q_e, Q_r, and Q_n.

Exercise 1.2 – Notation for doping reactions

A – MgO doped with Al_2O_3

1. Write the reaction for doping magnesium oxide MgO with aluminum oxide Al_2O_3 for the two following situations:
   a. the aluminum occupies an interstitial position,
   b. the aluminum substitutes for magnesium.
   For each case, give the electroneutrality equation.

2. Given that the cationic radii of magnesium and aluminum are 86 and 68 pm, respectively, which position (interstitial or substitution) would aluminum preferentially occupy?

Data

    MgO crystallizes in a NaCl-type cubic structure and exhibits a dominant Schottky-type intrinsic disorder.

B – NiO doped with Li_2O

Give the reaction for doping nickel monoxide NiO with lithium oxide Li_2O in the presence of oxygen, given that the solid solution obtained is a p-type semiconductor and that the lithium substitutes for the nickel. Write the electroneutrality equation.

Data

    Assume that cationic Frenkel disorder is dominant in NiO.

C – La_2O_3 doped with SrO

1. Give the reaction for doping La_2O_3 with strontium oxide SrO. Assume that La_2O_3 is dominated by anionic Frenkel disorder. Write the reactions that involve only atomic structure defects.

2. Assuming that Schottky disorder dominates in $LaMnO_3$,
   a. give the equilibrium relation associated with Schottky disorder in $LaMnO_3$ and the corresponding equilibrium constant.
   b. We study three different ways to introduce the dopant SrO into the $LaMnO_3$ structure. We consider that the solid solutions obtained take the form of an ideal cubic perovskite structure ABO_3. This structure is shown in figure 3.

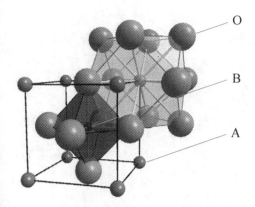

Figure 3 – Representation of unit cell of ideal perovskite structure ABO_3.

Strontium and lanthanum occupy the A sites and manganese occupies the B sites.

▷ Keeping the stoichiometry satisfied, complete each of the following reactions:

$$xSrO + LaMnO_3 \longrightarrow \ldots \tag{1}$$

$$xSrO + (1-x)LaMnO_3 \longrightarrow \ldots \tag{2}$$

$$xSrO + La_{1-x}MnO_{3-1.5x} \longrightarrow \ldots \tag{3}$$

▷ For each case, give the normal structure elements and the point defects.

▷ Calculate the reagent masses required to prepare 10 g of the solid solution $La_{0.84}Sr_{0.16}MnO_{2.92}$ from dehydrated lanthanum nitrate, manganese nitrate, and strontium carbonate.

Data

Elements	La	Sr	Mn	C	N	O
Molar mass [g mol⁻¹]	138.92	87.63	54.95	12	14	16

▷ Cite an application of this material.

c. Manganese is present in the solid solution $La_{1-x}Sr_xMnO_{3-x/2}$ in the ionic forms Mn^{2+}, Mn^{3+}, and Mn^{4+}.

▷ Use Kroger-Vink notation to express the disproportionation equilibrium involving these species.

▷ Express the equilibrium of the solid solution with gaseous oxygen involving only the manganese and the oxygen.

▷ By considering the atomic and electronic defects, give the expression that describes electric neutrality.

▷ Give the sitoneutrality relations for the formula $La_{1-x}Sr_xMnO_{3-x/2}$. Denote by [i] the site fraction occupied by species i, taking as reference sub-lattice A.

Exercise 1.3 – Sitoneutrality and expression of chemical formulas

The chemical formula describing a given solid solution of yttria-stabilized zirconia can be written in the two following manners:

$$(ZrO_2)_{1-x}(Y_2O_3)_x \quad \text{and} \quad (ZrO_2)_{1-y}(YO_{1.5})_y$$

1. Find the relationship $y = f(x)$.

2. Write the two formulas for $x = 0.08$.

3. For each case, calculate the number of cationic sites for one formula unit.

4. **a.** Write the formula $(ZrO_2)_{1-x}(Y_2O_3)_x$ in the form $Zr_\alpha Y_\beta O_\gamma V_\delta$, expressing α, β, γ, and δ as functions of x. V denotes an oxygen vacancy. Evaluate the result for $x = 0.08$.

   b. Write the formula $(ZrO_2)_{1-y}(YO_{1.5})_y$ in the form $Zr_\alpha Y_\beta O_\gamma V_\delta$, expressing α, β, γ, and δ as functions of y. Evaluate the result for $x = 0.08$.

Exercise 1.4 – Calculation of defect concentrations

Calculate the concentration C_v in mol L^{-1} of oxygen vacancies in the solid solution $(CeO_2)_{1-x}(CaO)_x$ with $x = 0.1$.

Data

> The solid solution crystallizes in a fluorite-type cubic structure.
> The lattice parameter is 5.415 Å.
> The dominant disorder in CeO_2 is of anionic Frenkel type and the calcium substitutes for cerium.

Exercise 1.5 – Doping strontium fluoride

Anionic Frenkel disorder is the dominant ionic disorder in strontium fluoride SrF_2.

1. What defects are present in a sample of SrF_2 doped simultaneously with potassium fluoride KF and lanthanum fluoride LaF_3? The dopant cations substitute for Sr^{2+} in the solid.

2. What fluoride dopant should be chosen to obtain an ionic conductor *via* a vacancy mechanism?

3. Consider preparing a solid solution by mixing 12.1832 g of SrF_2, 0.0581 g of KF, and 0.3918 g of LaF_3.
   a. Write the formula for this solid solution in the form $(SrF_2)_{1-\alpha}(KF)_\beta(LaF_3)_\gamma$. Give the expression relating α, β, and γ.
   b. What is the dominant defect in this solid solution?

   Data

Elements	Sr	K	La	F
Molar mass $[g\,mol^{-1}]$	87.6	39.1	138.9	19.0

4. Determine the molar concentration of vacancies (expressed as $mol\,cm^{-3}$) for 5% KF doping, given that the cubic-lattice parameter is 5.799 Å.

Exercise 1.6 – Variation of the concentration of structure defects in pure zirconium dioxide ZrO_2 as a function of oxygen partial pressure

Zirconium oxide ZrO_2 exhibits a dominant intrinsic Schottky-type disorder.

1. Write the reaction that involves the Schottky disorder in this oxide.

2. Establish the law for the variation in concentration of cationic defects with oxygen partial pressure for the range of very high oxygen partial pressure.

   Data

   No defect ionization is considered.

   Denote by K_S, K_g, and K_e the constants for the Schottky equilibrium, the equilibrium with the gaseous phase, and the electronic equilibrium, respectively.

   Consider the case $K_S \gg K_e$.

3. For the same domain, give the law for the variation in concentration of electrons as a function of oxygen partial pressure.

Exercise 1.7 – The non-stoichiometry of iron monoxide

Iron monoxide is a non-stoichiometric iron-deficient compound. A chemical analysis of a sample of iron monoxide gives a molar fraction of iron of 0.4767.

1. Give two formulas for the compound susceptible to account for this non-stoichiometry and comment on them with respect to a perfect FeO crystal.

2. Iron monoxide crystallizes in the cubic NaCl-type structure shown in figure 4.

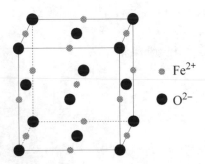

$\bullet$ Fe^{2+}

$\bullet$ O^{2-}

Figure 4 – Representation of NaCl-type structure of iron monoxide.

Table 2 gives the lattice parameter a and the density ρ of the sample analyzed.

a [Å]	4.371
ρ [g cm^{-3}]	5.312

Table 2 – Results of the analysis of a sample of iron monoxide.

What is the formula for this oxide?

Data

| Molar masses: $M(Fe) = 55.85$ g mol^{-1} $M(O) = 16$ g mol^{-1}

3. Write the equilibrium reaction involving this oxide and oxygen in the ambient atmosphere.

4. Explain the presence of the trivalent cation Fe^{3+} in the monoxide under study.

5. Write the formula for non-stoichiometric iron monoxide Fe$_{1-\alpha}$O in the form Fe$_x^{2+}$Fe$_y^{3+}$V$_z$O, with x, y, and z expressed as functions of α. V denotes an iron vacancy.

Exercise 1.8 – Departure from stoichiometry of barium fluoride BaF$_2$

At high temperature, solid barium fluoride BaF$_2$ exhibits an atomic disorder of the anionic Frenkel type. Consider a BaF$_2$ crystal placed in a container in which we can vary the partial pressure of fluorine gas F$_{2(g)}$. Assume that the entire system is at thermodynamic equilibrium.

1. Using Kröger-Vink notation, enumerate the various structure elements in the crystal. For each case, give the real charge of the corresponding center.

2. Draw the Brouwer diagram log[i] = f(log P_{F_2}) where [i] and P_{F_2} denote the concentration of species i and the partial pressure of fluorine, respectively. We shall consider successively the P_{F_2} domains corresponding to:
   a. high P_{F_2},
   b. low P_{F_2},
   c. Intermediate P_{F_2}. For this domain, draw the diagram for the two following cases:

   ▷ ionic defects are dominant

   Data
   $$\left| \quad K_{AF} = 10^{-8} \qquad K_e = 10^{-17} \qquad K_g = 10^{-23} \right.$$

   ▷ electronic defects are dominant

   Data
   $$\left| \quad K_{AF} = 10^{-17} \qquad K_e = 10^{-8} \qquad K_g = 10^{-23} \right.$$

   Note – Assume that the fluoride ions occupy normal and interstitial positions.

3. The departure from stoichiometry of BaF$_{2(1+\varepsilon)}$ can be expressed by the quantity δ, which is defined by $\delta = |F'_i| - |V_F^\bullet|$, where $|F'_i|$ and $|V_F^\bullet|$ are site fractions.
   a. Establish the expression relating δ to ε and $|F'_i|$ or $|V_F^\bullet|$. Assume that each crystal cell can accept only one interstitial anion.
   b. Establish the law describing how δ varies with fluorine partial pressure for the various domains of approximation given in question 2.

Exercise 1.9 – Crystallographic and thermodynamic study of thorium dioxide ThO$_2$

Thorium dioxide ThO$_2$ crystallizes as a cubic system. The symmetry of thorium dioxide is described by the space group Fm3m. The structure is composed of a face-centered-cubic lattice of Th^{4+} ions, with the tetrahedral sites occupied by O^{2-} ions. Anionic Frenkel-type disorder dominates in ThO$_2$.

A – Crystallographic study

1. Show the structure of ThO$_2$ in perspective and in standard [001] projection. Is the stacking compact for the face-centered-cubic (fcc) Th^{4+} ions?

2. Calculate the lattice parameter of thorium dioxide.

   Data
   $$\left| \; r(Th^{4+}) = 1.02 \text{ Å} \qquad r(O^{2-}) = 1.40 \text{ Å} \right.$$

3. Study the sites occupied by interstitial oxygen atoms.

B – Equilibrium with gas phase

1. **Pure ThO$_2$**
   a. Write the various equilibria and the relationships between the concentrations of structure defects, the equilibrium constants, and, where appropriate, the oxygen partial pressure.

      Remarks:

      ▷ write only the independent equilibria,

      ▷ use the following notation:
      - K$_e$ for the electronic equilibrium constant,
      - K$_g$ for the equilibrium constant for the gas phase,
      - K$_{AF}$ for the equilibrium constant for the formation of intrinsic defects.

      ▷ K$_g$ and K$_{AF}$ are defined for oxide ions in their normal position and with an activity equal to one.

   b. Establish the laws of variation in concentration of each type of defect as a function of oxygen partial pressure. Assimilate the activity to the concentration of each species. Consider successively the low-pressure, intermediate-pressure, and high-pressure domains by making the appropriate approximations and justifying them. In the intermediate-pressure domain, we treat only the case $K_e \ll K_{AF}$.

c. Using logarithmic coordinates, draw the corresponding curves (i.e., the Brouwer diagram).

2. Doped ThO_2

a. Consider ThO_2 doped with $YO_{1.5}$ at 5 at%, where the Y^{3+} ion occupies the thorium site. By applying the same reasoning as for question B1c, draw the Brouwer diagram for doped ThO_2. Denote by D_m the concentration of extrinsic defects.

b. What changes are caused by the doping?

c. What is the point of this doping?

Data

$$K_{AF} = 10^{-10} \qquad K_e = 10^{-16} \qquad K_g = 8.10^{-30}$$

Solutions to exercises

Solution 1.1 – Notation for structure elements and structure defects

1. The following expression relates the effective charge Q_e of a structure element or defect to the real charge Q_r of the ionic species that occupies a site and to the charge Q_n of the species that normally occupies the site:

$$Q_e = Q_r - Q_n$$

2. Listed below are the structure elements and defects written in the Kröger-Vink notation and with the values given for Q_e, Q_r, and Q_n:

Structure elements	Q_r	Q_n	Q_e	Notation
a. Sodium vacancy in NaCl	0	+1	−1	V'_{Na}
b. Fluoride (F^-) in normal position in CaF_2	−1	−1	0	F^x_F
c. Bromine vacancy in NaBr	0	−1	+1	$V^{\bullet}_{Br}$
d. Magnesium cation (Mg^{2+}) in normal position in MgO	+2	+2	0	Mg^x_{Mg}
e. Interstitial silver cation (Ag^+) in AgBr	+1	0	+1	$Ag^{\bullet}_i$
f. Oxygen vacancy in TiO_2	0	−2	+2	$V^{\bullet\bullet}_O$
g. Oxygen vacancy with one trapped electron in ZrO_2	−1	−2	+1	$V^{\bullet}_O$
h. Interstitial oxide ion with two trapped electron holes	0	0	0	O^x_i

3. Listed below are the structure elements and defects written in the Kröger-Vink notation and with the values given for Q_e, Q_r, and Q_n:

Structure elements	Q_r	Q_n	Q_e	Notation
a. Na^+ cation in substitution position of Mg^{2+} in MgO	+1	+2	−1	Na'_{Mg}
b. Oxide ion O^{2-} in substitution position of F^- in SrF_2	−2	−1	−1	O'_F
c. Al^{3+} ion in interstitial position in CaO	+3	0	+3	$Al^{3\bullet}_i$
d. Gd^{3+} ion in substitution position in CeO_2	+3	+4	−1	Gd'_{Ce}

4. Expressions for associated defect pairs written in Kröger-Vink notation and with values given for Q_e, Q_r, and Q_n:

Structure element	Q_r	Q_n	Q_e	Notation
a. Gd^{3+} ion in substitution position in CeO_2 and an oxygen vacancy	$+3$	$\begin{array}{c}+4-2\\=+2\end{array}$	$+1$	$(Gd_{Ce}V_O)^{\bullet}$
b. Chlorine and sodium vacancies in NaCl	0	$\begin{array}{c}-1+1\\=0\end{array}$	0	$(V_{Cl}V_{Na})^{\times}$

Solution 1.2 – Notation for doping reactions

A – MgO doped with Al_2O_3

1. For MgO, the equilibrium associated with Schottky disorder is expressed as:

$$0 \longrightarrow V''_{Mg} + V^{\bullet\bullet}_O$$

The reaction for doping MgO by Al_2O_3 is written as follows:

a. with Al in an interstitial position:

$$Al_2O_3 \longrightarrow 2Al^{3\bullet}_i + 3O^{\times}_O + 3V''_{Mg}$$

or $$Al_2O_3 + 3V^{\bullet\bullet}_O \longrightarrow 2Al^{3\bullet}_i + 3O^{\times}_O$$

The electroneutrality equation is

$$2[V^{\bullet\bullet}_O] + 3[Al^{3\bullet}_i] + p = 2[V''_{Mg}] + n$$

where n and p respectively denote the concentrations of electrons and holes.

b. with Al in the substitution position of magnesium:

$$Al_2O_3 \longrightarrow 2Al^{\bullet}_{Mg} + 3O^{\times}_O + V''_{Mg}$$

or $$Al_2O_3 + V^{\bullet\bullet}_O \longrightarrow 2Al^{\bullet}_{Mg} + 3O^{\times}_O$$

In this case, the electroneutrality equation is

$$2[V^{\bullet\bullet}_O] + [Al^{\bullet}_{Mg}] + p = 2[V''_{Mg}] + n$$

2. The cationic sub-lattice in MgO is face-centered-cubic (fcc), where the octahedral sites are occupied by oxygen, leading to a compact structure. Because the cationic radius of magnesium is close to that of aluminum, substitution is favored.

B – NiO doped with Li$_2$O

The equilibrium corresponding to the intrinsic disorder is

$$Ni^x_{Ni} \rightleftharpoons V''_{Ni} + Ni^{\bullet\bullet}_i$$

The reaction for doping NiO with Li$_2$O is

$$Li_2O + \tfrac{1}{2}O_2 \longrightarrow 2Li'_{Ni} + 2O^x_O + 2h^\bullet$$

The appearance of electron holes h$^\bullet$ gives rise to p-type semiconductivity.

The electroneutrality relation is

$$2[Ni^{\bullet\bullet}_i] + p = [Li'_{Ni}] + 2[V''_{Ni}] + n$$

Note – Some of the holes created are trapped by the nickel. Under these conditions, we can write

$$Li_2O + \tfrac{1}{2}O_2 + 2Ni^x_{Ni} \longrightarrow 2Li'_{Ni} + 2O^x_O + 2Ni^\bullet_{Ni}$$

Using x to denote the doping level gives

$$\tfrac{1}{2}xLi_2O + \tfrac{1}{4}xO_2 + (1-x)NiO \longrightarrow xLi'_{Ni} + O^x_O + (1-2x)Ni^x_{Ni} + xNi^\bullet_{Ni}$$

C – La$_2$O$_3$ doped with SrO

1. Consider the reaction for doping La$_2$O$_3$ with SrO.
   The equilibrium describing the dominant disorder in La$_2$O$_3$ is

$$O^x_O \rightleftharpoons V^{\bullet\bullet}_O + O''_i$$

   The reaction to introduce SrO is

$$SrO \longrightarrow Sr'_{La} + O^x_O + \tfrac{1}{2}V^{\bullet\bullet}_O$$

   or

$$xSrO + \tfrac{1}{2}(1-x)La_2O_3 \longrightarrow xSr'_{La} + (1-x)La^x_{La} + \tfrac{1}{2}(3-x)O^x_O + \tfrac{1}{2}xV^{\bullet\bullet}_O$$

   The electroneutrality relation is

$$2[V^{\bullet\bullet}_O] + p = [Sr'_{La}] + 2[O''_i] + n$$

2. Consider the reaction to dope LaMnO$_3$ with SrO
   a. The Schottky equilibrium in LaMnO$_3$ and the corresponding equilibrium constant are written as

$$0 \rightleftharpoons V^{3'}_{La} + V^{3'}_{Mn} + 3V^{\bullet\bullet}_O \; ; \quad K_S = [V^{3'}_{La}] [V^{3'}_{Mn}] [V^{\bullet\bullet}_O]^3$$

b. ▷ The reactions are

$$xSrO + LaMnO_3 \longrightarrow LaSr_xMnO_{3+x} \qquad (1)$$

$$xSrO + (1-x)\,LaMnO_3 \longrightarrow La_{1-x}Sr_xMn_{1-x}O_{3-2x} \qquad (2)$$

$$xSrO + La_{1-x}MnO_{3-1.5x} \longrightarrow La_{1-x}Sr_xMnO_{3-0.5x} \qquad (3)$$

▷ The normal structure elements and the point defects in the solid solution are given below:

$$xSrO + LaMnO_3 \longrightarrow La_{La}^{x} + x\,Sr_{La}' + (3+x)O_O^{x}$$
$$+ Mn_{Mn}^{x} + x\,V_{Mn}^{3'} + 2x\,V_O^{\bullet\bullet} \qquad (1)$$

$$x\,SrO + (1-x)LaMnO_3 \longrightarrow (1-x)La_{La}^{x} + x\,Sr_{La}' + (1-x)Mn_{Mn}^{x}$$
$$+ (3-2x)O_O^{x} + xV_{Mn}^{3'} + 2xV_O^{\bullet\bullet} \qquad (2)$$

$$x\,SrO + La_{1-x}MnO_{3-1.5x} \longrightarrow (1-x)La_{La}^{x} + x\,Sr_{La}' + Mn_{Mn}^{x}$$
$$+ (3-0.5x)O_O^{x} + 0.5xV_O^{\bullet\bullet} \qquad (3)$$

▷ Preparation of the solid solution $La_{0.84}Sr_{0.16}MnO_{2.92}$:

Calculation of the molar mass of the solid solution to prepare

$$M = (0.84 \times M_{La}) + (0.16 \times M_{Sr}) + M_{Mn} + (2.92 \times M_O)$$

$$M = 232.38 \text{ g mol}^{-1}$$

10 g of this solid solution corresponds to an amount of substance n of

$$n = \frac{10}{232.38}$$

$$n = 0.043 \text{ mol}$$

We are left to calculate the molar masses of the reagents and the masses to weigh given the stoichiometric coefficients of each element in the solid solution. The molar masses and the quantity of reagent and product are listed in table 3.

Table 3 – Molar masses and quantity of reagents and product

Parameters	$La(NO_3)_3$	$SrCO_3$	$Mn(NO_3)_3$	$La_{0.84}Sr_{0.16}MnO_{2.92}$
Molar mass [g mol^{-1}]	324.92	147.63	240.95	232.38
n [mol per 10 g]	0.0361	0.00688	0.043	0.043
m [g per 10 g]	11.73	1.016	10.36	10.00

▷ This material is used for cathodes in solid oxide fuel cells (SOFCs).

c. ▷ Disproportionation equilibrium in manganese is written as

$$2Mn^{\times}_{Mn} \rightleftarrows Mn'_{Mn} + Mn^{\bullet}_{Mn}$$

▷ With gaseous oxygen, we can consider the two following equilibria:

$$\tfrac{1}{2}O_{2(g)} + 2Mn^{\times}_{Mn} + V^{\bullet\bullet}_O \rightleftarrows O^{\times}_O + 2Mn^{\bullet}_{Mn}$$

$$\tfrac{1}{2}O_{2(g)} + 2Mn'_{Mn} + V^{\bullet\bullet}_O \rightleftarrows O^{\times}_O + 2Mn^{\times}_{Mn}$$

The linear combination of the two preceding reactions allows us to find the disproportionation equilibrium of Mn^{3+}.

Note – The presence of the redox couple $Mn^{\bullet}_{Mn}/Mn^{\times}_{Mn}$ (Mn^{4+}/Mn^{3+}) involved in the equilibrium

$$Mn^{\times}_{Mn} + h^{\bullet} \rightleftarrows Mn^{\bullet}_{Mn}$$

is, at high oxygen partial pressure, the source of the p-type semiconductivity of this material.

▷ The relation describing electric neutrality is

$$2[V^{\bullet\bullet}_O] + [Mn^{\bullet}_{Mn}] + p = [Sr'_{La}] + [Mn'_{Mn}]$$
$$+ 3[V^{3'}_{La}] + 3[V^{3'}_{Mn}] + n$$

▷ Sitoneutrality relations:
Denote by [i] the fraction of sites occupied by species i with the lanthanum sub-lattice taken as reference.

  ▷ for lanthanum: $[La^{\times}_{La}] + [Sr'_{La}] + [V^{3'}_{La}] = 1$
  ▷ for manganese: $[Mn'_{Mn}] + [Mn^{\times}_{Mn}] + [Mn^{\bullet}_{Mn}] + [V^{3'}_{Mn}] = 1$
  ▷ for oxygen: $[O^{\times}_O] + [V^{\bullet\bullet}_O] = 3$

Solution 1.3 – Sitoneutrality and notation for chemical formulas

1. The fraction of sites occupied by zirconium must be the same independent of notation:

$$\frac{1-x}{1-x+2x} = \frac{1-x}{1+x} \qquad \text{for } (ZrO_2)_{1-x}(Y_2O_3)_x$$

and

$$\frac{1-y}{1-y+y} = 1-y \qquad \text{for } (ZrO_2)_{1-y}(YO_{1.5})_y$$

or $$1 - y = \frac{1 - x}{1 + x}$$

which gives $$y = \frac{2x}{1 + x}$$

2. For $x = 0.08$, we have $y = 0.148$. The formulas for the solid solution are

$$(ZrO_2)_{0.92}(Y_2O_3)_{0.08} \quad \text{and} \quad (ZrO_2)_{0.852}(YO_{1.5})_{0.148}$$

3. For a formula unit, the number of cationic sites in each case is

for $(ZrO_2)_{0.92}(Y_2O_3)_{0.08}$ $n_{Zr} = 0.92 + 2 \times 0.08$

or $n_{Zr} = 1.08$

for $(ZrO_2)_{0.852}(YO_{1.5})_{0.148}$ $n_{Zr} = 0.852 + 0.148$

or $n_{Zr} = 1$

4. **a.** Expression of the formula $(ZrO_2)_{1-x}(Y_2O_3)_x$ in the form $Zr_\alpha Y_\beta O_\gamma V_\delta$ with α, β, γ, and δ expressed as a function of x.
   We first transform the formula into the form $Zr_{1-x}Y_{2x}O_{2(1-x)+3x}V_\delta$
   By identification, we have $\alpha = 1 - x$, $\beta = 2x$, and $\gamma = 2 + x$
   For δ, the number of anionic sites is twice that of cationic sites, from which we deduce

   $$\delta + 2(1-x) + 3x = 2(1+x)$$

   or $$\delta = x$$

   Thus, $(ZrO_2)_{1-x}(Y_2O_3)_x$ may be written in the form

   $$Zr_{1-x}Y_{2x}O_{2+x}V_x$$

   The numerical application to $(ZrO_2)_{1-x}(Y_2O_3)_x$ for $x = 0.08$ gives

   $$Zr_{0.92}Y_{0.16}O_{2.08}V_{0.08}$$

   b. Expression of the formula $(ZrO_2)_{1-y}(YO_{1.5})_y$ in the form $Zr_\alpha Y_\beta O_\gamma V_\delta$ with α, β, γ and δ expressed as a function of x
   We transform the formula into the form $Zr_{1-y}Y_yO_{2(1-y)+\frac{3}{2}y}V_\delta$.
   By identification, we have $\alpha = 1 - y$, $\beta = y$, and $\gamma = 2 - \frac{1}{2}y$.

For δ, the number of anionic sites is twice that of cationic sites, from which we deduce

$$\delta + 2 - \tfrac{1}{2}\,y = 2$$

or
$$\delta = \tfrac{1}{2}y$$

Thus, $(ZrO_2)_{1-y}(YO_{1.5})_y$ may be written in the form

$$Zr_{1-y}Y_yO_{2-\frac{1}{2}y}V_{\frac{1}{2}y}$$

Numerical evaluation of $(ZrO_2)_{1-y}(YO_{1.5})_y$ for $x = 0.08$ gives

$$Zr_{0.852}Y_{0.148}O_{1.926}V_{0.074}$$

Solution 1.4 – Calculation of defect concentrations

We begin by writing the reaction for doping CeO_2 by CaO:

$$CaO \longrightarrow Ca''_{Ce} + O_O^\times + V_O^{\bullet\bullet}$$

or
$$xCaO + (1-x)CeO_2 \longrightarrow xCa''_{Ce} + (1-x)Ce_{Ce}^\times + (2-x)O_O^\times + xV_O^{\bullet\bullet}$$

Note that the concentration of oxygen vacancies is the same as that of Ca''_{Ce}. The fluorite-type structure is shown in figure 5.

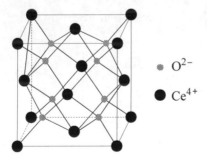

$$\circledast \quad O^{2-}$$

$$\bullet \quad Ce^{4+}$$

Figure 5 – Schematic diagram of the ideal structure of ceria.

A unit cell contains four Ce sites and eight O sites. The number of sites occupied by vacancies is 4x per unit cell, which gives a vacancy concentration of

$$C_V = \frac{4x}{N_{Av} \times V_m}$$

where N_{Av} and V_m denote Avogadro constant number and the unit-cell volume, respectively. Numerical evaluation gives

$$C_V = \frac{4 \times 0.1}{(6.02 \times 10^{23})(5.415 \times 10^{-8})^3} = 4.185 \times 10^{-3}\ \mathrm{mol\,cm^{-3}}$$

or
$$C_V = 4.185 \, \text{mol L}^{-1}$$

This is a high concentration compared with the usual aqueous solutions.

Solution 1.5 – Doping strontium fluoride

1. Given that anionic Frenkel disorder is the dominant ionic disorder in stron-tium fluoride SrF_2, we have the following equilibrium:

$$F_F^x \rightleftharpoons V_F^\bullet + F_i'$$

The equations for introducing dopants are

$$KF \longrightarrow K_{Sr}' + F_F^x + V_F^\bullet$$

$$LaF_3 \longrightarrow La_{Sr}^\bullet + 2F_F^x + F_i'$$

For simultaneous doping, the reaction is obtained by combining the two preceding individual reactions:

$$KF + LaF_3 \longrightarrow K_{Sr}' + La_{Sr}^\bullet + 3F_F^x + V_F^\bullet + F_i'$$

The structure defects in the final phase are

$$V_F^\bullet, \ F_i', \ K_{Sr}', \ \text{and} \ La_{Sr}^\bullet$$

2. For conduction by ionic vacancy to be the dominant mechanism, we must use KF as dopant because it will increase the concentration of anionic vacancies.

3. **a.** The molar masses and amounts of various constituents required for making a solid solution are gathered in table 4.

Table 4 – Molar masses and amount
of various constituents required for making solid solution

Parameters	SrF_2	KF	LaF_3
Molar mass [g mol^{-1}]	125.6	58.1	195.9
m [g]	12.1832	0.0581	0.3918
n [mol]	0.097	0.001	0.002

Based on table 4, we obtain 0.1 mol of solid solution. We can thus deduce that, for one mole, the chemical formula for a solid solution is

$$(SrF_2)_{0.97} \, (KF)_{0.01} \, (LaF_3)_{0.02} \qquad \text{with } \alpha = \beta + \gamma$$

b. The equation of mixed doping is

$$0.01KF + 0.02LaF_3 \longrightarrow 0.01K'_{Sr} + 0.02La^{\bullet}_{Sr} + 0.06F^X_F + 0.01F'_i$$

Note that the dominant defect is interstitial fluorine.

4. The chemical formula for the solid solution with 5% doping is $(SrF_2)_{0.95}(KF)_{0.05}$ or $Sr_{0.95}K_{0.05}F_{1.95}V_{0.05}$, where V represents the fluorine vacancies. Note that the number of vacancies equals the number of potassium cations. The molar concentration of vacancies may be calculated for a unit cell as follows:

$$C = \frac{n_{V,m}}{V_m} = \frac{4 \times x}{N_{Av}V_m}$$

where x is the doping level, N_{Av} is Avogadro constant, V_m is the volume of the unit cell, and $n_{V,m}$ is the amount of substance in the volume V_m.

We obtain
$$C = \frac{4 \times 0.05}{(6.02 \times 10^{23}) \times (5.799 \times 10^{-8})^3}$$

or
$$C = 1.7 \times 10^{-3} \text{ mol cm}^{-3}$$

Solution 1.6 – Variation of the concentration of structure defects in pure zirconium dioxide ZrO_2 as a function of oxygen partial pressure

1. For ZrO_2, the Schottky equilibrium is expressed as

$$0 \rightleftharpoons V^{4'}_{Zr} + 2V^{\bullet\bullet}_O$$

where the constant K_S is

$$K_S = [V^{4'}_{Zr}][V^{\bullet\bullet}_O]^2$$

2. The concentration of cationic defects as a function of oxygen partial pressure $[V^{4'}_{Zr}] = f(P_{O_2})$.

To derive this relationship, we must express

▷ the equilibrium with gaseous oxygen

$$O^X_O \rightleftharpoons \tfrac{1}{2}O_2 + 2e' + V^{\bullet\bullet}_O$$

$$K_g = [V^{\bullet\bullet}_O]\, n^2\, P^{1/2}_{O_2}$$

▷ the electronic equilibrium

$$0 \rightleftharpoons e' + h^{\bullet}$$

$$K_e = np$$

▷ and the electroneutrality relation

$$p + 2[V_O^{\bullet\bullet}] = n + 4[V_{Zr}^{4\prime}]$$

At high oxygen partial pressure, the species $V_{Zr}^{4\prime}$ and $h^{\bullet}$ dominate, which gives the following approximation:

$$p = 4[V_{Zr}^{4\prime}]$$

from which

$$K_g^2 = [V_O^{\bullet\bullet}]^2 \times n^4 \times P_{O_2} = [V_O^{\bullet\bullet}]^2 \times \frac{K_e^4}{p^4} \times P_{O_2} = \frac{K_S}{[V_{Zr}^{4\prime}]} \times \frac{K_e^4}{4^4[V_{Zr}^{4\prime}]^4} \times P_{O_2}$$

$$[V_{Zr}^{4\prime}] = 0.33 \times \left(\frac{K_S K_e^4}{K_g^2}\right)^{1/5} P_{O_2}^{1/5}$$

3. The variation in the concentration n of electrons as a function of oxygen partial pressure $n = f(P_{O_2})$ is

$$n = \frac{K_e}{p} = \frac{K_e}{4[V_{Zr}^{4\prime}]} = \frac{K_e}{4} \times \frac{1}{0.33} \times \left(\frac{K_g^2}{K_S K_e^4}\right)^{1/5} P_{O_2}^{-1/5}$$

$$n = 0.76 \times \left(\frac{K_g^2 K_e}{K_S}\right)^{1/5} P_{O_2}^{-1/5}$$

Solution 1.7 – The non-stoichiometry of iron monoxide

1. The problem statement gives a molar fraction of iron in the oxide of 0.4767. The molar fraction of oxygen is thus

$$x_O = 1 - x_{Fe}$$

$$x_O = 1 - 0.4767 = 0.5233$$

We deduce the two following formulas for the monoxide:

$$Fe_{0.91}O \quad \text{and} \quad FeO_{1.098}$$

With respect to a perfect FeO crystal, the first formula indicates an iron deficiency and the second indicates an excess of oxygen.

2. There are $Z = 4$ formula units FeO in the unit cell. If FeO is stoichiometric, the density ρ_{th} is

$$\rho_{th} = \frac{ZM_{FeO}}{N_{Av}a^3}$$

where M_{FeO} is the molar mass of the oxide.

The calculation gives $\rho_{th} = \dfrac{4 \times (55.85 + 16)}{(6.02 \times 10^{23}) \times (4.371 \times 10^{-8})^3}$

$$\rho_{th} = 5.717 \text{ g cm}^{-3}$$

Because the measured density ($\rho = 5.312$ g cm^{-3}) is less than that of the stoichiometric monoxide, we conclude that the formula $Fe_{0.91}O$ should be retained with an incompletely filled iron sub-lattice.

3. The equilibrium reaction involving the monoxide and oxygen is

$$\tfrac{1}{2}O_{2(g)} \rightleftharpoons O_O^\times + 2h^\bullet + V_{Fe}''$$

4. The presence of trivalent cations Fe^{3+} may be explained by hole trapping by Fe^{2+} cations according to the reaction

$$Fe_{Fe}^\times + h^\bullet \rightleftharpoons Fe_{Fe}^\bullet$$

Note – The equilibrium reaction with oxygen may also be written as

$$2Fe_{Fe}^\times + \tfrac{1}{2}O_{2(g)} \rightleftharpoons O_O^\times + 2Fe_{Fe}^\bullet + V_{Fe}''$$

5. Based on one mole of $Fe_{1-\alpha}O$, electroneutrality, mass balance, and sitoneutrality lead respectively to

$$2x + 3y - 2 = 0$$
$$x + y = 1 - \alpha$$
$$x + y + z = 1$$

We deduce
$$x = 1 - 3\alpha$$
$$y = 2\alpha$$
$$z = \alpha$$

Finally, the formula for the monoxide is

$$Fe_{1-3\alpha}^{2+}Fe_{2\alpha}^{3+}V_\alpha O$$

Solution 1.8 – Departure from stoichiometry of barium fluoride BaF$_2$

1. Given that the dominant atomic disorder in BaF$_2$ is anionic Frenkel disorder, the various structure elements are

Structure element	Ba$_{Ba}^{\times}$	F$_F^{\times}$	F$_i'$	V$_F^{\bullet}$	V$_i^{\times}$
Real charge of center Q$_r$	+2	−1	−1	0	0

Note – V$_i^{\times}$ denotes an empty interstitial site. In general, it does not appear in the list of structure elements present in a crystalline phase.

2. To draw the Brouwer diagram, one must express the various equilibria between the ionic and electronic defects and the electroneutrality relation.

▷ Equilibria involving the ionic and electronic defects (table 5)

Table 5 – Equilibria considered in BaF$_2$ and the corresponding constants.

Reaction	Law of mass action	Simplified relation	
Equilibrium involving ionic defects:			
$F_F^{\times} + V_i^{\times} \rightleftharpoons F_i' + V_F^{\bullet}$	$K_{AF} = \dfrac{[F_i'][V_F^{\bullet}]}{[F_F^{\times}][V_i^{\times}]}$	$K_{AF} = [F_i'][V_F^{\bullet}]$	(1)
Equilibrium involving gaseous P$_{F_2}$:			
$\frac{1}{2}F_{2(g)} + V_i^{\times} \rightleftharpoons F_i' + h^{\bullet}$	$K_g = \dfrac{[F_i']p}{P_{F_2}^{1/2}[V_i^{\times}]}$	$K_g = \dfrac{[F_i']p}{P_{F_2}^{1/2}}$	(2)
Equilibrium involving electronic defects:			
$0 \rightleftharpoons e' + h^{\bullet}$	$K_e = np$	$K_e = np$	(3)

▷ Electroneutrality relation

$$[V_F^{\bullet}] + p = [F_i'] + n \qquad (4)$$

This relation must be simplified by retaining only a single term from each member.

a. Case of high fluorine partial pressure

By considering the preceding equilibria, we find that p and [F$_i'$] dominate, which gives
$$p = [F_i']$$

b. Case of low fluorine partial pressure

In this case, n and [V$_F^{\bullet}$] dominate, so
$$[V_F^{\bullet}] = n$$

c. Case of intermediate fluorine partial pressure with
   ▷ ionic defects dominate ($K_{AF} \gg K_e$).
    Under these conditions, we have $[V_F^\bullet] = [F_i']$
   ▷ electronic defects dominate ($K_e \gg K_{AF}$).
    Here we have the approximation $p = n$

The solution of the system of equations (1)–(4) gives the variation in concentration of each defect as a function of P_{F_2}. The results are gathered in table 6.

Table 6 – Variation of defect concentration as a function of fluorine pressure.

Defect	Low pressure	Intermediate pressure		High pressure
		$K_{AF} \gg K_e$ ionic defects dominate	$K_{AF} \ll K_e$ electronic defects dominate	
$[F_i']$	$\propto P_{F_2}^{1/4}$	$= [V_F^\bullet] = K_{AF}^{1/2} = $ const.	$\propto P_{F_2}^{1/2}$	$\propto P_{F_2}^{1/4}$
p	$\propto P_{F_2}^{1/4}$	$\propto P_{F_2}^{1/2}$	$= n = K_e^{1/2} = $ const.	$\propto P_{F_2}^{1/4}$
$[V_F^\bullet]$	$\propto P_{F_2}^{-1/4}$	$= [F_i'] = K_{AF}^{1/2} = $ const.	$\propto P_{F_2}^{-1/2}$	$\propto P_{F_2}^{-1/4}$
n	$\propto P_{F_2}^{-1/4}$	$\propto P_{F_2}^{-1/2}$	$= p = K_e^{1/2} = $ const.	$\propto P_{F_2}^{-1/4}$

Figure 6 shows the Brouwer diagram corresponding to defects in BaF_2.

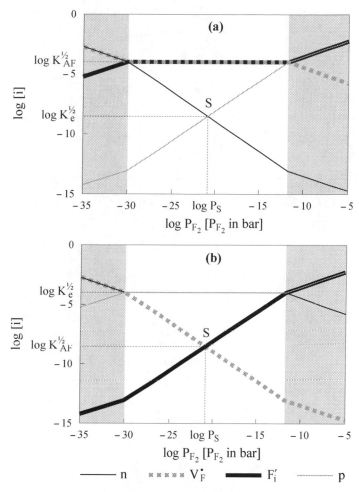

Figure 6 – Brouwer diagram for BaF_2 with **(a)** $K_{AF} \gg K_e$, and **(b)** $K_{AF} \ll K_e$. P_S is the pressure at the stoichiometric point S.

3. a. To derive the expression relating δ to ε, we must express ε as a function of the site fractions $|F_i'|$ and $|V_F^{\bullet}|$. The formula $BaF_{2(1+\varepsilon)}$ dictates

$$\frac{\text{total number of F}}{\text{total number of Ba}} = 2(1 + \varepsilon)$$

or

$$\frac{N_{F_F^{\times}} + N_{F_i'}}{N_{Ba_{Ba}^{\times}}} = 2(1 + \varepsilon) \qquad (5)$$

Moreover, we have

$$N_{Ba_{Ba}^{\times}} = \tfrac{1}{2} N_{\text{fluorine sites}} = 4 N_{\text{interstitial sites}}$$

or $\qquad \dfrac{N_{F_F^\times}}{N_{Ba_{Ba}^\times}} = 2\,|F_F^\times| \quad$ and $\quad \dfrac{N_{F_i'}}{N_{Ba_{Ba}^\times}} = \dfrac{|F_i'|}{4}$

By using equation (1), we deduce

$$|F_F^\times| + \frac{|F_i'|}{8} = 1 + \varepsilon \tag{6}$$

On the other hand, we have

$$|F_F^\times| + |V_F^\bullet| = 1 \tag{7}$$

and $\qquad\qquad\qquad\qquad |F_i'| - |V_F^\bullet| = \delta \tag{8}$

By combining equations (6)–(8), we eliminate two types of site fractions and obtain

$$\delta = \tfrac{7}{8}\,|F_i'| + \varepsilon \quad \text{and} \quad \delta = 7\,|V_F^\bullet| + 8\,\varepsilon$$

b. The variation in δ as a function of fluorine pressure

We use the expression $\delta = |F_i'| - |V_F^\bullet|$, which we simplify based on the dominant species in the various ranges of fluorine pressure.

▷ High pressures

$$|F_i'| \gg |V_F^\bullet| \longrightarrow \delta \approx |F_i'| \longrightarrow \delta \propto P_{F_2}^{1/4}$$

▷ Low pressures

$$|V_F^\bullet| \gg |F_i'| \longrightarrow \delta \approx |V_F^\bullet| \longrightarrow \delta \propto P_{F_2}^{-1/4}$$

▷ Intermediate pressures

▷ Ionic defects dominate ($K_{AF} \gg K_e$)

$$|F_i'| \approx |V_F^\bullet| \longrightarrow \delta \approx 0$$

▷ Electronic defects dominate ($K_e \gg K_{AF}$)

$$|F_i'| \propto P_{F_2}^{1/2} \quad \text{and} \quad |V_F^\bullet| \propto P_{F_2}^{-1/2}$$

which gives $\qquad \delta \approx a\,P_{F_2}^{1/2} - b\,P_{F_2}^{-1/2}$

where a and b are constants.

Solution 1.9 – Crystallographic and thermodynamic study of thorium dioxide ThO₂

A – Crystallographic study

1. The ideal crystalline structure for ThO₂ is shown in figure 7 (a) in perspective and (b) in standard [001] projection.

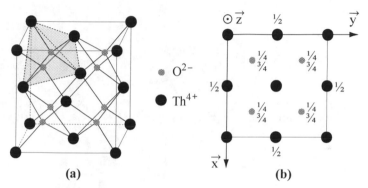

Figure 7 – Representation of the ideal structure of ThO₂
(a) in perspective and (b) in standard [001] projection.

For compact stacking, the Th^{4+} ions are joined along a face diagonal (fig. 8), which gives (with a the lattice parameter of ThO₂):

$$4r_{Th^{4+}} = a \times \sqrt{2} \quad \text{and} \quad a = \frac{4r_{Th^{4+}}}{\sqrt{2}}$$

$$a = \frac{4 \times 1.02}{\sqrt{2}}$$

$$a = 2.885 \text{ Å}$$

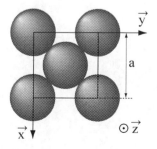

Figure 8 – Representation of the cationic sub-lattice of ThO₂ for the case of compact stacking of Th^{4+} ions.

Under these conditions, does the face-centered-cubic lattice provide sufficient space for the O^{2-} ions to occupy the tetrahedral sites?

As shown in figures 7(a) and 9, the tetrahedral sites are regular and aligned along [111] rows (i.e., the $\bar{3}$ axes and the diagonals of the cubic cell).

Consequently,
$$r_{Th^{4+}} + r_{Td} = \frac{a\sqrt{3}}{4}$$

where r_{Td} is the maximum radius of an atom at a tetrahedral site in the compact stacking formed by the Th^{4+} ions.

$$r_{Td} = \frac{a\sqrt{3}}{4} - r_{Th^{4+}}$$

$$r_{Td} = \frac{2.885\sqrt{3}}{4} - 1.02$$

$$r_{Td} = 0.23 \, \mathring{A} < r_{O^{2-}}$$

The tetrahedral site cannot accommodate the O^{2-} ions. Consequently, the stacking of the Th^{4+} ions is not compact.

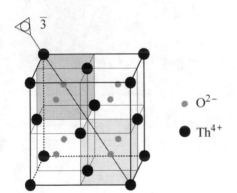

$\cdot$ O^{2-}

$\bullet$ Th^{4+}

Figure 9 – Structural representation of tetrahedral sites (Td) occupied by O^{2-} ions in the stacking of the Th^{4+} ions.

2. Calculation of the lattice parameter of ThO_2

The same reasoning as above leads to
$$\frac{a\sqrt{3}}{4} = r_{Th^{4+}} + r_{O^{2-}}$$

from which
$$a = \frac{4}{\sqrt{3}}(r_{Th^{4+}} + r_{O^{2-}})$$

$$a = \frac{4}{\sqrt{3}}(1.02 + 1.4)$$

$$a = 5.589 \, \mathring{A}$$

3. Interstitial hosting sites for oxide ions

As shown in figure 10, the ThO_2 structure has four octahedral sites per unit cell, denoted Oh, at the center of the unit cell and in the middle of the edges.

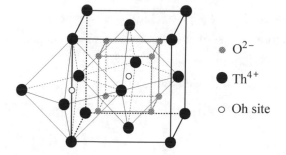

• O^{2-}

● Th^{4+}

o Oh site

Figure 10 – Structural representation of octahedral sites (Oh) in ThO_2.

Considering the Th^{4+} ions to be the nearest neighbors of the octahedral site, we have

$$\frac{a}{2} = r_{Th^{4+}} + r_{Oh}$$

from which

$$r_{Oh} = \tfrac{1}{2}(a - 2r_{Th^{4+}})$$

$$r_{Oh} = \tfrac{1}{2}(5.589 - 2 \times 1.02)$$

$$r_{Oh} = 1.77 \, \text{Å} > r_{O^{2-}}$$

Inserting O^{2-} ions in octahedral sites is possible in a fcc lattice of Th^{4+} ions. Considering the O^{2-} ions to be nearest neighbors of the octahedral sites, we have

$$\frac{a\sqrt{3}}{2} = 2r_{O^{2-}} + r_{Oh}$$

or

$$r_{Oh} = \frac{1}{2}\left(\frac{a\sqrt{3}}{2} - 2r_{O^{2-}}\right)$$

$$r_{Oh} = \frac{1}{2}\left[\frac{5.589 \times \sqrt{3}}{2} - (2 \times 1.4)\right]$$

We deduce that the maximum radius of an atom occupying an octahedral site without deforming the lattice is

$$r_{Oh} = 1.02 \, \text{Å} < r_{O^{2-}}$$

Consequently, insertion of an O^{2-} ion into an octahedral site must be accompanied by a deformation of the ThO_2 host lattice.

B – Equilibrium with the gaseous phase

1. Pure ThO$_2$

a. Notation for the various equilibria and the relations existing between the concentrations of structural point defects, the equilibrium constants, and, where appropriate, the oxygen partial pressure

▷ Equilibrium involving the intrinsic atomic disorder

$$O_O^x \rightleftharpoons V_O^{\bullet\bullet} + O_i'' \qquad\qquad K_{AF} = [V_O^{\bullet\bullet}][O_i'']$$

▷ Electronic equilibrium

$$0 \rightleftharpoons e' + h^\bullet \qquad\qquad K_e = n\,p$$

▷ Equilibrium with the gaseous phase

$$O_O^x \rightleftharpoons \tfrac{1}{2}O_{2(g)} + V_O^{\bullet\bullet} + 2e' \qquad K_g = [V_O^{\bullet\bullet}]n^2\,P_{O_2}^{1/2}$$

b. The law of variation in the concentration of each species as a function of oxygen partial pressure

To establish these laws, we first have to write the electroneutrality relations

$$n + 2[O_i''] = p + 2[V_O^{\bullet\bullet}]$$

This relation can be simplified for each range of oxygen partial pressure by keeping only one term from each member (Brouwer approximation). When $K_e \ll K_{AF}$, this gives the following diagram:

$$n = 2[V_O^{\bullet\bullet}] \qquad\qquad [O_i''] = [V_O^{\bullet\bullet}] \qquad\qquad p = 2[O_i'']$$

$$\xrightarrow{\hspace{8cm}} \log P_{O_2}$$

At low oxygen partial pressure, the Brouwer approximation gives

$$n = 2[V_O^{\bullet\bullet}]$$

from which

$$K_g = \tfrac{1}{2}n^3\,P_{O_2}^{1/2}$$

and

$$n = 2^{1/3} K_g^{1/3} P^{-1/6}$$

where P denotes oxygen partial pressure.

Table 7 lists the laws of variation in the concentration of other defects obtained for each range of approximation.

Table 7 – Laws of variation in concentrations of defects in pure ThO_2 as a function of oxygen partial pressure P (Brouwer approximation).

P_{O_2}	n	$[V_O^{\bullet\bullet}]$	$[O_i'']$	p
low $n = 2[V_O^{\bullet\bullet}]$	$2^{1/3}K_g^{1/3}P^{-1/6}$	$4^{-1/3}K_g^{1/3}P^{-1/6}$	$4^{1/3}K_g^{-1/3}K_{AF}P^{1/6}$	$2^{-1/3}K_g^{-1/3}K_eP^{1/6}$
intermediate $[O_i''] = [V_O^{\bullet\bullet}]$	$K_g^{1/2}K_{AF}^{-1/4}P^{-1/4}$	$K_{AF}^{1/2}$	$K_{AF}^{1/2}$	$K_g^{-1/2}K_{AF}^{1/4}K_eP^{1/4}$
high $p = 2[O_i'']$	$2^{-1/3}K_e^{1/3}K_g^{1/3}$ $K_{AF}^{-1/3}P^{-1/6}$	$4^{1/3}K_g^{1/3}K_{AF}^{2/3}$ $K_e^{-2/3}P^{-1/6}$	$4^{-1/3}K_g^{-1/3}K_{AF}^{1/3}$ $K_e^{2/3}P^{1/6}$	$2^{1/3}K_e^{2/3}K_g^{-1/3}$ $K_{AF}^{1/3}P^{1/6}$

c. Drawing a Brouwer diagram

In a Brouwer diagram, the variation in concentrations of defect [i] takes the from of a linear equation: $\log[i] = A \times \log P + B$.

The coefficients A and B are obtained based on the relations found in table 7 and on the values of the constants K_{AF}, K_e, and K_g. The results are gathered in tables 8 and 9 for each defect and for each range of partial pressure.

Table 8 – Ordinate values (on a log scale) at the origin of the lines in the Brouwer diagram for each charge carrier and for three ranges of oxygen partial pressure.

P_{O_2} range	n	$[V_O^{\bullet\bullet}]$	$[O_i'']$	p
Low P_{O_2}	−9.60	−9.90	−0.10	−6.40
Intermediate P_{O_2}	−12.05	−5.00	−5.00	−3.95
High P_{O_2}	−11.80	−5.50	−4.50	−4.20

Table 9 – Slopes of the lines in the Brouwer diagram for each charge carrier and for three ranges of oxygen partial pressure.

P_{O_2} range	n	$[V_O^{\bullet\bullet}]$	$[O_i'']$	p
Low P_{O_2}	−0.167	−0.167	0.167	0.167
Intermediate P_{O_2}	−0.250	0	0	0.250
High P_{O_2}	−0.167	−0.167	0.167	0.167

The Brouwer diagram for ThO_2 is shown in figure 11(a).

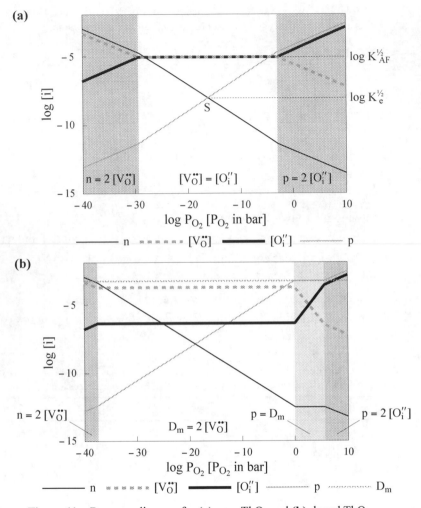

Figure 11 – Brouwer diagram for **(a)** pure ThO_2 and **(b)** doped ThO_2.

2. Doped ThO_2

a. The equilibria for the intrinsic atomic and electronic disorders as well as the equilibria with the gas phase remain unchanged.

The doping reaction is

$$Y_2O_3 \longrightarrow 2Y'_{Zr} + 3O^\times_O + V^{\bullet\bullet}_O$$

The electroneutrality relation becomes

$$n + 2[O''_i] + [Y'_{Zr}] = p + 2[V^{\bullet\bullet}_O]$$

Hereafter, $[Y'_{Zr}]$ is denoted D_m.

As before, this relation can be simplified. For doped ThO_2, when $K_e \ll K_{AF}$, we are led to consider a fourth range.

The diagram for the approximation becomes

$$n = 2[V_O^{\bullet\bullet}] \quad D_m = 2[V_O^{\bullet\bullet}] \qquad\qquad p = D_m \qquad p = 2[O_i''] \longrightarrow \log P_{O_2}$$

Applying a reasoning identical to that of the previous question leads to the laws of variation in defect concentration for each range of approximation. These laws are gathered in table 10.

Table 10 – Laws of variation in defect concentration for doped ThO_2 as a function of oxygen partial pressure P (Brouwer approximation).

P_{O_2}	n	$[V_O^{\bullet\bullet}]$	$[O_i'']$	p
Low $n = 2[V_O^{\bullet\bullet}]$	$2^{1/3}K_g^{1/3}P^{-1/6}$	$4^{-1/3}K_g^{1/3}P^{-1/6}$	$4^{1/3}K_g^{-1/3}K_{AF}P^{1/6}$	$2^{-1/3}K_g^{-1/3}K_eP^{1/6}$
Intermediate $D_m = 2[V_O^{\bullet\bullet}]$	$2^{1/2}D_m^{-1/2}K_g^{1/2}P^{-1/4}$	$\dfrac{D_m}{2}$	$\dfrac{2K_{AF}}{D_m}$	$2^{-1/2}D_m^{1/2}K_eK_G^{-1/2}P^{1/4}$
High $p = D_m$	$\dfrac{K_e}{D_m}$	$K_GK_e^{-2}D_m^2P^{-1/2}$	$\dfrac{K_{AF}K_G^{-1}K_e^2}{D_m^{-2}P^{1/2}}$	D_m
Very high $p = 2[O_i'']$	$2^{-1/3}K_e^{1/3}K_g^{1/3}K_{AF}^{-1/3}P^{-1/6}$	$4^{1/3}K_g^{1/3}K_{AF}^{2/3}K_e^{-2/3}P^{-1/6}$	$4^{-1/3}K_g^{-1/3}K_{AF}^{2/3}K_e^{2/3}P^{1/6}$	$2^{1/3}K_e^{2/3}K_g^{-1/3}K_{AF}^{1/3}P^{1/6}$

D_m remains constant over the entire range of pressure. The Brouwer diagram for doped ThO_2 appears in figure 11(b).

b. The consequences of doping are

  ▷ an increase in the oxygen vacancy concentration, leading to an increase in ionic conductivity,

  ▷ an increase in the electrolytic domain, where ionic conductivity dominates,

  ▷ the appearance of a fourth domain where the concentration of extrinsic defects is dominant with respect to the concentration of interstitial oxygen.

Note – Note that the concentration of vacancies introduced by the doping is considerably greater than the concentration of intrinsic vacancies.

   c. The point of such a doping strategy is to increase the ionic conductivity
and the range of ionicity in order to obtain a stable solid electrolyte over
a large range of oxygen pressure.

Chapter 2

Methods and techniques

Course notes
───────

2.1 – Complex impedance spectroscopy

Complex impedance spectroscopy measures the response of an electrochemical system to a small-amplitude ac voltage perturbation of variable frequency around a steady-state operational point.

2.1.1 – Time domain: principal passive linear dipole devices in sinusoidal regime

We denote voltage and current by v(t) and i(t), respectively. The current-voltage characteristics of the principal dipole devices in the time domain are

For a resistance R
$$v(t) \; = \; R \; i(t)$$

For a capacitance C
$$v(t) \; = \; \frac{1}{C} \int_t i(t)\,dt$$

For an inductance L
$$v(t) \; = \; L\frac{di(t)}{dt}$$

2.1.2 – Complex notation

The instantaneous complex expressions for $\overline{v}$ and $\overline{i}$ of the voltage v(t) and the current i(t) respectively derive from the real expressions

$$v(t) \; = \; V \sin(\omega t + \alpha)$$
$$\overline{v} \; = \; V \, e^{j\alpha} \, e^{j\omega t}$$
$$i(t) \; = \; I \sin(\omega t + \beta)$$
$$\overline{i} \; = \; I \, e^{j\beta} \, e^{j\omega t}$$

© Springer Nature Switzerland AG 2020
A. Hammou and S. Georges, *Solid-State Electrochemistry*,
https://doi.org/10.1007/978-3-030-39659-6_2

Complex notation allows us to express the current-voltage characteristics in the sinusoidal regime in the form $\quad \overline{v} = Z \times \overline{i}$

where Z is the complex impedance.

The complex impedance and its inverse (complex admittance Y) are written respectively as

$$Z(\omega) = \frac{\overline{v}(t)}{\overline{i}(t)} = \frac{V\,e^{j\omega t + \Phi}}{I\,e^{j\omega t}} = R + jX = |Z|\,e^{j\Phi}$$

and $\qquad\qquad Y(\omega) = \frac{1}{Z(\omega)} = G + jB = |Y|\,e^{-j\Phi}$

where j is the imaginary unit with $j^2 = -1$, $Z(\omega)$ is the complex impedance, $Y(\omega)$ is the complex admittance, R is the real part of the impedance, called the resistance, X is the imaginary part of the impedance, called the reactance, G is the real part of the admittance, called the conductance, B is the imaginary part of the admittance, called the susceptance, and $\Phi = \alpha - \beta$ is the phase shift of the current i(t) with respect to the voltage v(t) (i.e., the phase difference).

Applying the complex notation to the three basic dipole devices allows us to express their complex impedance by calculating the ratio $\overline{v}(t)/\overline{i}(t)$.

For a resistance $\qquad\qquad\qquad\qquad Z_R(\omega) = R$

For a capacitance $\quad Z_C(\omega) = \dfrac{\overline{v}(t)}{\overline{i}(t)} = \dfrac{1}{jC\omega} = \dfrac{1}{C\omega}e^{-j\frac{\pi}{2}}$

For an inductance $\quad Z_L(\omega) = \dfrac{\overline{v}(t)}{\overline{i}(t)} = jL\omega = L\omega\,e^{j\frac{\pi}{2}}$

By applying the rules for combining these three elements,

$$Z_{tot}(\omega) = \sum_k Z_k(\omega) \text{ in series and } \frac{1}{Z_{tot}(\omega)} = \sum_k \frac{1}{Z_k(\omega)} \text{ in parallel,}$$

we directly obtain the expression for the complex impedance of any arbitrary electric circuit.

2.1.3 – Graphical representation of complex impedance

The expression for the impedance of a R//C circuit is given by applying the rules for combining these elements to the expressions of the complex impedance of the various dipole devices

$$Z_{(RC)}(\omega) = \frac{R}{1 + R^2 C^2 \omega^2} - j\left[\frac{R^2 C\omega}{1 + R^2 C^2 \omega^2}\right]$$

Similarly, k R//C circuits connected in series have a total impedance

$$Z_{(RC)_k}(\omega) = \sum_k \left[\frac{R_k}{1 + R_k^2 C_k^2 \omega^2} \right] - j\sum_k \left[\frac{R_k^2 C_k \omega}{1 + R_k^2 C_k^2 \omega^2} \right]$$

The representation used by electrochemists plots the opposite of the imaginary part $Z'' = -\text{Im}[Z(\omega)]$ as a function of the real part $Z' = \text{Re}[Z(\omega)]$ of the complex impedance. The diagram obtained is called a Nyquist diagram. Figure 12 shows a Nyquist diagram for a R//C circuit.

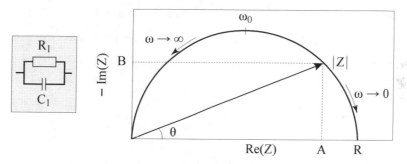

Figure 12 – Nyquist representation of a R//C circuit.

The notable values obtained from this diagram are

▷ the intersection of the circle with the real axis, which gives the total resistance of the system

$$\lim_{\omega \to \infty}\{\text{Re}[Z(\omega)]\} = 0 \quad \text{and} \quad \lim_{\omega \to 0}\{\text{Re}[Z(\omega)]\} = R$$

▷ the modulus $|Z|$ of the complex impedance

$$|Z| = \sqrt{A^2 + B^2}$$

with $A = |Z|\cos\theta$ and $B = |Z|\sin\theta$

▷ the argument of the complex impedance $\theta = \arctan\dfrac{B}{A}$

▷ the relaxation frequency $\quad f_0 = \dfrac{1}{2\pi RC} = \dfrac{\omega_0}{2\pi}$

The Nyquist representation does not give the frequency for each point on the diagram. It is often useful to consider other complementary representations, called Bode plots. These are shown in figure 13 for the case of two R//C circuits in series.

Figure 13 shows

$\triangleright$ a Nyquist representation $-\operatorname{Im}[Z(\omega)] = f\{\operatorname{Re}[Z(\omega)]\}$ **(a)**

$\triangleright$ a Bode representation $-\operatorname{Im}[Z(\omega)] = f[\log(\omega)]$ **(b)**

$$\operatorname{Re}[Z(\omega)] = f[\log(\omega)]$$ **(c)**

and $$\theta = f[\log(\omega)]$$ **(d)**

As shown in figure 14, two semicircles appear in these plots if the ratio of relaxation frequencies (f_0) is sufficiently large (of the order of five).

These four representations allow us to visualize several characteristic quantities of an electric circuit. The same representations are possible with the admittance Y, which gives the conductance G and the susceptance B.

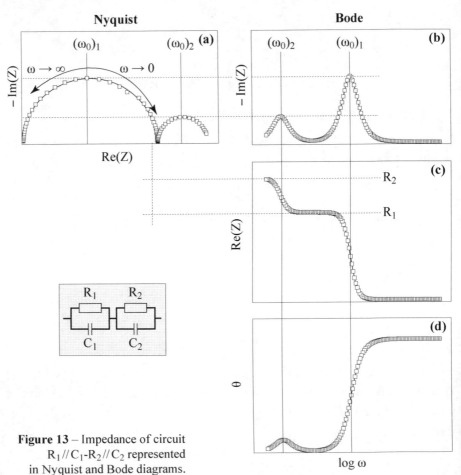

Figure 13 – Impedance of circuit $R_1 /\!/ C_1$-$R_2 /\!/ C_2$ represented in Nyquist and Bode diagrams.

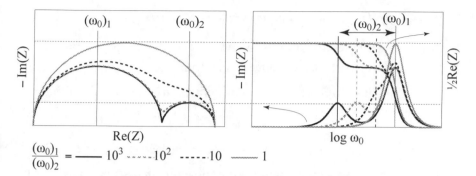

Figure 14 – Representations in the Nyquist and Bode planes of the impedance of $R_1//C_1$-$R_2//C_2$ circuits for various ratios of relaxation frequencies.

2.1.4 – Other dipole devices

The constant phase element (CPE) is a dipole device with two parameters: a pseudo-capacitance A (expressed in $F\,s^{p-1}$) and an exponent p. Its electrical behavior cannot be reproduced by combining basic R, C, L elements. The complex impedance of a CPE is

$$\underset{CPE}{Z}(\omega) = \frac{1}{A(j\omega)^p}$$

The exponent determines the phase angle β: $\beta = \dfrac{p\pi}{2}$ with $0 \le p \le 1$. It gives a real (resistive) component to the CPE; in other words, a resistance (for all p different from unity), that is not exhibited by the classic capacitance, as shown by the following relation:

$$\underset{CPE}{Z}(\omega) = \frac{A\omega^p \cos\dfrac{p\pi}{2}}{\left(A\omega^p \cos\dfrac{p\pi}{2}\right)^2 + \left(A\omega^p \sin\dfrac{p\pi}{2}\right)^2}$$
$$- j\left[\frac{A\omega^p \sin\dfrac{p\pi}{2}}{\left(A\omega^p \cos\dfrac{p\pi}{2}\right)^2 + \left(A\omega^p \sin\dfrac{p\pi}{2}\right)^2}\right]$$

Depending on p, the CPE behaves like a pure dipole device:

▷ $p = 1$, $Z(\omega) = 1/A(j\omega)$ (pure capacitance with C = A)

▷ $p = 0$, $Z(\omega) = 1/A$ (pure resistance with R = 1/A)

▷ $p = -1$, $Z(\omega) = A(j\omega)$ (pure inductance with L = A)

The relaxation frequency ω_0 of a R//CPE circuit is

$$\omega_0 = \frac{1}{(RA)^{1/p}}$$

2.1.5 – Physical meaning of complex impedance spectra

By analogy with electric circuits, complex impedance spectra provide the following information for a conducting material:

▷ $Re(Z)_{max} = R$: sample resistance. Its conductivity σ is obtained from

$$\sigma = \frac{1}{R} \times \frac{\ell}{S}$$

where ℓ and S are the sample thickness and surface area, respectively.

▷ $Im(Z)_{max}$ is the maximum phase shift. The equivalent capacitance C is obtained from

$$C = \varepsilon_0 \varepsilon_R \frac{S}{\ell}$$

where ε_0 is the vacuum permittivity ($\varepsilon_0 = 8.8542 \times 10^{-12}\ F\,m^{-1}$) and ε_R is the dielectric constant (or relative permittivity) of the sample.

The impedance spectra of ceramics generally show three domains, as shown in figure 15. Phenomena associated with grains appear at high frequencies, those associated with the microstructure (grain boundaries) appear at intermediate frequencies, and those associated with the electrode reaction appear at low frequencies.

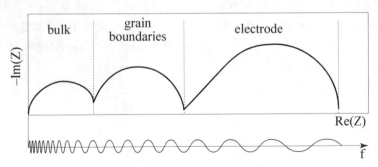

Figure 15 – General form of impedance spectrum for a ceramic.

2.2 – Methods to measure transport number

2.2.1 – Electromotive force method

The presentation of this method is based on the mixed conductor (ionic + electronic) described by the formula MX (M^+X^-). Consider the following electrochemical chain:

$$Me/\mu_X^{(1)}|MX|\mu_X^{(2)}/Me$$

MX is inserted between two metallic electrodes Me (1) and (2) where the chemical potential μ_X of species X is fixed. The following equilibrium holds in each electrode:

$$X + e \rightleftharpoons X^-$$

Given the potential difference ΔE measured at the terminals of the chain, we can determine the average ionic transport number $\bar{t}_i$ in the domain $\mu_X^{(1)} - \mu_X^{(2)}$ by applying the Wagner model

$$\bar{t}_i = \frac{\Delta E}{\Delta E_{th}}$$

ΔE_{th} is the theoretical electromotive force (emf) that we calculate by applying the Nernst equation:

$$\Delta E_{th} = \frac{\mu_X^{(1)} - \mu_X^{(2)}}{F}$$

When the electrode electrochemical equilibria are ensured by fixing the partial pressures $P_{X_2}^{(1)}$ and $P_{X_2}^{(2)}$ at the electrodes, ΔE_{th} is given by

$$\Delta E_{th} = \frac{RT}{2F} \ln \frac{P_{X_2}^{(1)}}{P_{X_2}^{(2)}}$$

The measurement of ΔE can be distorted by the existence of an electrochemical semipermeability flux, which, in particular, polarizes the electrodes and/or introduces a slow mechanism of adsorption-desorption of the gaseous species at the electrodes.

Note – An identical approach can be developed for the case where a chemical-potential gradient is imposed on the metallic species M at the electrodes.

2.2.2 – Using the results of total conductivity

This method exploits the variation of the total conductivity σ_t of a mixed conductor MX as a function of the partial pressure of X_2. The range of pressure under study must be sufficiently large to lead to the appearance of

▷ a plateau of constant ionic conduction σ_i ($t_i = 1$) and

▷ a zone where the total conductivity varies. Under these conditions, we have

$$\sigma_t = \sigma_i + \sigma_e$$

from which we deduce the electronic transport number t_e

$$t_e = \frac{\sigma_e}{\sigma_t}$$

Note that this method implies that the ionic conductivity remains constant throughout the range over which total conductivity varies. Given the large difference between ionic and electronic mobilities, this is a reasonable hypothesis, particularly when the species responsible for ionic transport is extrinsic (e.g., doping).

Finally, this method is above all applicable to mixed conductors with non-negligible electronic transport number.

2.2.3 – Tubandt method

This method is based on that of Hittorf, which is used in the electrochemistry of solutions. Schematically, it is based on exploiting the consecutive mass balances of an electrolysis in a (cathode or anode) compartment, taking into account

▷ the formation or disappearance of the ionic species present in the electrolyte following the electrode reaction, and,

▷ in the compartment under consideration, the arrival or departure by migration of the ionic species.

When applied to an MX compound, the method consists of measuring the amount of substance that appears or disappears in one (or several) compartments upon considering the following electrode reaction:

$$\tfrac{1}{2}X_2 + e \rightleftharpoons X^-$$

This quantity is directly related to the cationic transport number. Figure 16 shows schematically the principle of the method.

The measurement cell consists of three MX tablets (1), (0), and (2) with two inert electrodes at the extremities. It is traversed by 1 Faraday and, for simplicity, we consider only compartment (1). Through migration, it loses t_c moles of M^+ and gains t_a moles of X^-. One mole of X^- disappears at the anode. The result is that compartment (1) loses t_c moles of MX. This loss $\Delta m = |m_1' - m_1|$, where m_1 and m_1' denote the masses of MX before and after electrolysis, respectively, is determined by weighing. We find that compartment (2) gains the same amount. The mass of the middle part (0) does not change. The cationic transport number is

$$t_c = \frac{F\Delta m}{MI\tau}$$

where I is the intensity of the electrolysis current, M is the molar mass of MX, and τ is the electrolysis time.

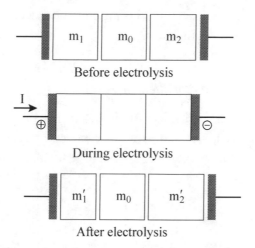

Figure 16 – Schematic of principle for measuring cationic transport number by using the Tubandt method (from Hladik, 1972).

2.2.4 – Dilatocoulometric method to measure cationic transport number

The principle is based on measuring the displacement of the electrode-electrolyte interfaces due to variations in mass in the cathode and anode compartments, itself caused by the electrolysis of the crystal under study (fig. 17).

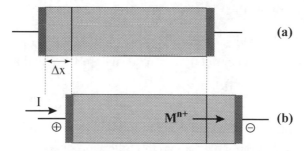

Figure 17 – Schematic diagram of dilatocoulometry device (a) before electrolysis and (b) after electrolysis.

The displacement Δx is followed by dilatometry. The method may be applied to simple ionic crystals with the formula M_aX_b. The anodic reaction should lead to a gaseous species (e.g., X_2). We show that the cationic transport number t_c is given by

$$t_c = \frac{zFS\rho\Delta x}{MI\tau}$$

where z is the charge of the cation, S is the electrode-electrolyte contact area, ρ is the crystal density, I is the electrolysis current, M is the molar mass of the crystal, and τ is the electrolysis time.

2.2.5 – Electrochemical semipermeability

Consider for example a solid oxide solution, such as stabilized zirconia, where the oxide ions give the ionic conductivity. At zero current and under a chemical potential gradient, a flux of oxygen crosses the material, compensated by a flux of electronic species (fig. 18).

$$P_2 > P_1$$

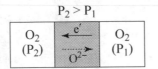

Figure 18 – Schematic representation of phenomenon of electronic semipermeability.

We thus have a non-equilibrium system that evolves over time. The fluxes are related by
$$J_e = -2J_{O^{2-}} = -4J_{O_2}$$

Under low oxygen partial pressure, we can write

$$J_e = -\tilde{u}RT|\vec{\nabla}[e']| = -\tilde{u}RT\frac{[e']^{(2)} - [e']^{(1)}}{L}$$

where L is the sample thickness, oriented from (1) toward (2).

At low oxygen partial pressure ($P_1 \ll P_2$), this reduces to $J_e = \tilde{u}RT\frac{[e']^{(1)}}{L}$.

Given that the electron conductivity is $\sigma_e = \tilde{u}F^2[e']^{(1)}$ and that $J_e = -4J_{O_2}$,

we obtain
$$J_{O_2} = \frac{RT}{4F^2L}\sigma_e^{(1)} \quad \text{and} \quad \sigma_e^{(1)} = \frac{4F^2L}{RT}J_{O_2}$$

At high partial pressure or where the hole conductivity σ_h dominates, we follow the same approach as above to obtain the following relation:

$$J_{O_2} = \frac{RT}{4F^2L}\sigma_h^{(2)} \quad \text{and} \quad \sigma_h^{(2)} = \frac{4F^2L}{RT}J_{O_2}$$

where σ_h represents the hole conductivity.

Note that the measurement of the oxygen flux allows us to calculate the electronic conductivity σ_e or σ_h.

In the oxygen partial pressure domain where the concentration of no electronic charge carrier dominates, we obtain

$$J_{O_2} = \frac{RT}{4F^2L}\left(\sigma_e^{(1)} + \sigma_h^{(2)}\right)$$

Experimentally (fig. 19), we measure the specific oxygen flux J_{O_2} given by

$$J_{O_2} = \frac{D}{V_m} \times \frac{\Delta P}{P_e} \times \frac{L}{S}$$

where D is the gas flow, V_m is the molar volume of the gas, ΔP is the variation in pressure due to the enrichment of the most-oxygen-deficient gas, P_e is the total pressure of the gas at the entrance, L is the sample thickness, and S is the surface area of the sample.

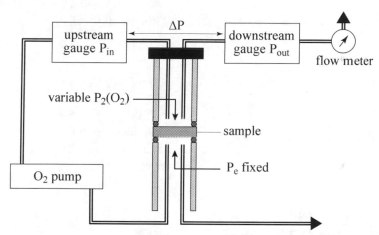

Figure 19 – Schematic diagram of experimental cell for measuring flux specific to electrochemical semipermeability.

Determining the electronic transport numbers t_e and t_h requires measuring the total conductivity and calculating the electronic conductivities from the flux due to electrochemical semipermeability.

2.2.6 – Blocking electrode method

The principle is based on measuring, in a mixed ionic-electronic conductor (MIEC),

▷ the ionic conduction alone by using an electron-blocking electrode made of a purely ionic conductor, or

▷ the electronic conduction alone by using an ion-blocking electrode made of a purely electronic conductor.

Figure 20 shows a schematic describing the principle of the Hebb-Wagner method applied to a MIEC conductor by oxide ions.

The electron-blocking electrode consists of a micro-point made of yttria-stabilized zirconia. The interface with the sample constitutes the working electrode WE. Applying a voltage U between the counter-electrode CE and the working electrode WE allows the local control of oxygen activity at the working electrode. Thus, the oxygen activity is fixed in the vicinity of the micro-electrode provided that the MIEC/micro-electrode interface is isolated from the gas phase by a tight glass seal. The activity is

$$a_{O_2}^{WE} = a_{O_2}^{RE} e^{\frac{4UF}{RT}}$$

Given the geometry of the system, the ionic conductivity is

$$\sigma_i\left(a_{O_2}^{WE}\right) = \frac{1}{2\pi a} \times \frac{dI}{dU}$$

where a is the contact radius of the micro-point in contact with the sample.

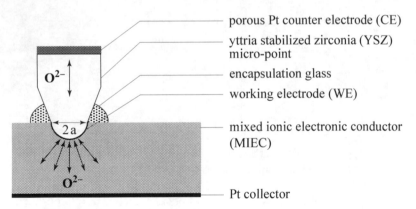

Figure 20 – Principle of measurement of ionic conductivity in a MIEC by using Hebb-Wagner method (from Zipprich & Wiemhöfer, 2000).

The electronic conductivity can be measured in a similar way with a platinum micro-point.

Exercises

Exercise 2.1 – Determination of conductivity by four-electrode method

The four-electrode method, which is depicted schematically in figure 21, is used to determine the electrical conductivity of the electrode material $La_{0.8}Sr_{0.2}MnO_{3-\delta}$ in air at various temperatures.

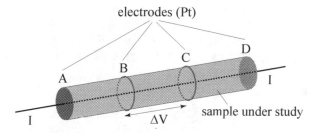

Figure 21 – Schematic diagram showing principle of four-electrode method.

The direct current is imposed in galvanostatic mode. The results obtained at 666 °C are listed in table 11.

Table 11 – Current at various potential differences between points B and C.

I [mA]	50	100	150	200	250	300	400	500
$V_B - V_C$ [mV]	2.11	4.22	6.33	8.44	10.55	12.66	16.88	21.09

1. Show that the sample obeys Ohm's law in the range of potential under study.

2. Determine the resistance $R_{B-C} = R_1$ between points B and C.

3. Given that the sample is cylindrical with a geometric factor $k = 7.59$ cm^{-1}, what is the conductivity of $La_{0.8}Sr_{0.2}MnO_{3-\delta}$ at 666 °C?

4. Based on the resistance R_{B-C} obtained at different temperatures (table 12),
   a. plot the conductivity σ for $La_{0.8}Sr_{0.2}MnO_{3-\delta}$ in the form $\log \sigma T = f(1/T)$;
   b. deduce the activation energy in eV.

Table 12 – Resistance R_{B-C} at various temperatures.

T [°C]	102	200	298	398	490	589	666	741
R [Ω]	0.1082	0.06938	0.05614	0.04954	0.04578	0.04357	0.04219	0.04125

5. The resistance R_{A-B}, where point A represents the metal (Pt) electrode, is 0.0142 Ω. The distance between points A and B (B and C) is 5 (1.496) mm. What can we conclude about
   a. the polarization resistance at the $La_{0.8}Sr_{0.2}MnO_{3-\delta}/Pt$ interface?
   b. the type of conduction in $La_{0.8}Sr_{0.2}MnO_{3-\delta}$ in air?

Exercise 2.2 – Measurement of electric quantities by complex impedance spectroscopy

We use complex impedance spectroscopy (CIS) to study the properties of oxide ion conductors. The sample consists of a cylindrical tablet 0.177 cm thick and with a surface area of 1.267 cm². Each face of the sample is covered with a platinum current-collecting electrode. The electrical measurements are done in air at 300 °C, in a frequency range spanning from 5 Hz to 13 MHz and with an ac voltage of 50 mV.

To determine the electric parameters required to characterize the material under study, the sample can be considered as an equivalent circuit.

We study two types of circuits: a circuit consisting of a resistance in parallel with a capacitance (R//C) and a circuit consisting of a resistance in parallel with a constant phase element (CPE) (R//CPE).

The complex impedance of a CPE is

$$Z_{CPE}(\omega) = \frac{1}{A(j\omega)^p}$$

where A is a pseudo-capacitance and ω is the frequency of the electric field. p is a decentering parameter (in the Nyquist representation) related to the distribution of relaxation times associated with the phenomena of ionic polarization in ceramics.

Adjusting the circuit parameters by a least squares fit gives the characteristics listed in table 13.

Table 13 – Electric characteristics determined at 300 °C.

Circuit	R [Ω]	C [F]	A [F s^{p-1}]	p
R//C	6.07×10^4	2.93×10^{-11}	–	–
R//CPE	6.19×10^4	–	1.41×10^{-10}	0.879

Figure 22 shows the experimental and calculated points (upper plots) of the normalized impedance and the residues (lower plots). k is the geometric factor of the sample.

The residues are obtained by taking the difference between the experimental and calculated points at each frequency.

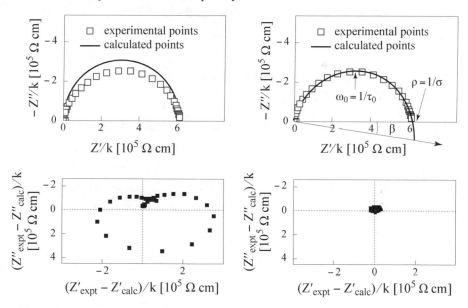

Figure 22 – Normalized experimental and calculated impedance spectra for two types of equivalent electric circuits and an analysis of the residues (left: R//C; right: R//CPE).

1. Derive the expression for total impedance $Z_{tot}(\omega)$ of a R//C circuit and that for a R//CPE circuit as a function of frequency ω of the electric field and in the form of a complex number $a + jb$. What does the expression for $Z_{R//CPE}(\omega)$ give for $p = 1$?

2. In the Nyquist plane, show the impedance of a R//CPE circuit with R = 0.8 MΩ, A = 8×10^{-11} F s^{p-1} for p = 1, p = 0.75, and p = 0.5 from 10^7 to 10^{-2} Hz. Show the decentering angle for p = 0.75 and p = 0.5.

3. From the given data (table 13), calculate
   a. the geometric factor k of the sample,
   b. the relaxation frequency ω_0,
   c. the time constant τ_0,
   d. the capacitance C,
   e. the decentering angle β,
   f. the dielectric constant ε_R,
   g. the permittivity ε,
   h. the electrical conductivity σ,
   i. the electrical resistivity ρ of the sample.

4. Of these parameters, which are independent of the sample shape?

5. What is the most appropriate equivalent circuit?

Exercise 2.3 – Measurement of electronic conductivity in a mixed conductor

The electronic conductivity of ceria doped with gadolinium oxide and with praseodymium oxide ($Ce_{0.8}Gd_{0.17}Pr_{0.03}O_{1.9}$, denoted CGP) is studied at 750 °C as a function of oxygen activity. The sample is inserted between an O^{2-}-ion-blocking platinum working electrode (WE) and a reference electrode (RE) to create a CuO-Cu_2O pair. Figure 23 shows a schematic diagram of the setup.

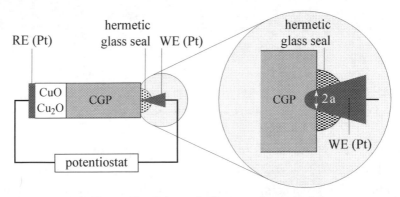

Figure 23 – Schematic diagram of setup and electrochemical cell (from Lübke & Wiemhöfer,1999).

The working electrode in contact with the sample has a radius of 170 μm. The WE/CGP interface is isolated from the gas phase by a tight glass seal.

The oxygen activity $a_{O_2}^{RE}$ at the reference electrode is fixed by the equilibrium between the two copper oxides

$$2CuO(s) \rightleftharpoons Cu_2O(s) + \tfrac{1}{2}O_2(g)$$

The working electrode is polarized in order to fix the oxygen activity at the WE-sample interface. The steady-state potential-current pairs (U, I) are given in table 14.

Table 14 – Potential-current pairs obtained in steady state.

U [V]	−0.5	−0.4	−0.3	−0.2	−0.1	0	0.1	0.2	0.3
I [μA]	−11	−4.8	−2.2	−1	−0.4	0	0.6	2.4	9.7

Based on the Hebb-Wagner model, the electronic conductivity $\sigma_e(a_{O_2}^{WE})$ is obtained from the slope of the I(U) curve:

$$\sigma_e(a_{O_2}^{WE}) = \frac{1}{2\pi a} \times \frac{dI}{dU}$$

where a is the radius of the WE-sample contact.

1. Make a diagram of the electrochemical chain.

2. Write the electrode reactions.

3. Derive an expression for the potential difference ΔE at the terminals of this chain as a function of $a_{O_2}^{WE}$ and $a_{O_2}^{RE}$. Deduce from this the expression for $a_{O_2}^{WE}$.

4. a. Calculate the values of $\log a_{O_2}^{WE}$ for each potential applied.
      Give an example and put the results in a table.
   b. Draw the curves for I(U) and $\log a_{O_2}^{WE}(U)$.
      Determine the equation for the curve $\log a_{O_2}^{WE}(U)$.
   c. Calculate the values for $\log a_{O_2}^{WE}$ and $\log \sigma_e(a_{O_2}^{WE})$ between two successive potentials. Give an example and put the results in a table for each average potential.

5. Represent the variation in electronic conductivity of CGP as a function of oxygen activity on a logarithmic scale and identify the branches σ_n and σ_p in the form

$$\log \sigma = \alpha \times \log a_{O_2}^{WE} + \beta$$

6. Calculate σ_n for $a_{O_2}^{WE} = 10^{-12}$ and σ_p for $a_{O_2}^{WE} = 1$.

Data

$$\left| \; a_{O_2}^{RE} = 2.68 \times 10^{-4} \right.$$

Exercise 2.4 – Measurement of ionic conductivity in a mixed conductor

We now study the ionic conductivity of strontium-doped lanthanum cobaltite $La_{0.8}Sr_{0.2}CoO_{3-\delta}$, (LSCo) at 600 °C as a function of oxygen activity. The ionic conductivity is measured by using a micro-point made of yttria-stabilized zirconia (YSZ), which is electronically insulating and serves as electrolyte. The LSCo sample studied in the form of a tablet constitutes the working electrode (WE). The contact radius (denoted a) between the zirconia and the sample is 0.01 cm. The counter electrode (CE) is a layer of porous platinum deposited on the face opposite the micro-point. The interface between the micro-point and the sample is isolated from the gas phase by a tight glass seal. The entire cell is placed in air, which has an oxygen partial pressure of 0.2 bar. Figure 24 shows a schematic diagram of the setup.

The sample is polarized to localize the oxygen activity at the micro-point interface. The potential-current pairs (U,I), measured in the steady state, are given in table 15.

Table 15 – Applied potentials and currents measured in steady state.

U [V]	−0.160	−0.140	−0.120	−0.100	−0.080	−0.060	−0.040	−0.020
I [nA]	−23	−16.2	−10.8	−7.15	−4.6	−2.8	−1.57	−0.71
U [V]	0	0.020	0.040	0.060	0.080	0.100	0.120	0.140	0.160
I [nA]	0	0.62	1.2	1.75	2.34	3.05	3.9	5.1	6.8

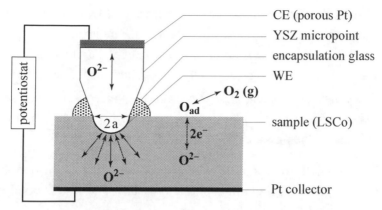

Figure 24 – Schematic diagram of electrochemical cell
(from Zipprich & Wiemhöfer, 2000).

Based on the Hebb-Wagner model, the ionic conductivity σ_i is obtained from the slope of the polarization curve I(U) as follows:

$$\sigma_i(a_{O_2}^{WE}) = \frac{1}{2\pi a} \times \frac{dI}{dU}$$

where a is the radius of the contact between the micro-point and sample.

1. Draw a diagram of the relevant electrochemical chain.

2. Derive the expression for the oxygen activity $a_{O_2}^{WE}$ at the working electrode as a function of the potential difference U between the chain terminals.

3. Draw the curve I(U) and determine an analytic expression for this function. Calculate the oxygen activity $a_{O_2}^{WE}$ and the ionic conductivity σ_i of the sample for each applied potential.

4. Using logarithmic coordinates, graph the ionic conductivity of the sample as a function of the oxygen activity and explain the variations observed.

Exercise 2.5 – Determination of cationic transport number by dilatocoulometry

1. Based on a simple diagram, give the principle of dilatocoulometry applied to an ionic binary compound MX (M^{z+}, X^{z-}) with a non-negligible cationic conductivity.

2. Derive the following relation: $t_c = zF \times \dfrac{\rho}{M} \times \dfrac{S\Delta x}{I\tau}$

where t_c is the cationic transport number in the MX compound, F is the Faraday constant, ρ is the density of MX, M is the molar mass of MX, S is the surface area of the anode, Δx is the displacement of the anode-MX interface, I is the dc current passing through the anode, and τ is the electrolysis time.

Consider the displacement of a cylindrical cell with constant cross section s and whose anodic interface has been displaced a distance Δx. This displacement is due to the passage of a current I for a time τ.

3. Dilatocoulometry was applied to determine the cationic transport number in silver iodide α-AgI at atmospheric pressure. The results obtained are presented in table 16.

Table 16 – Results obtained by dilatocoulometry for silver iodide α-AgI.

T [°C]	Δx [μm]	$I \times \tau$ [C]
250	48	1.76
300	50.50	1.80

a. State the requisite temperature conditions for this study.
b. Calculate the cationic transport number for α-AgI at each temperature.
c. What conclusion can we draw?

Data

$\left|\ \rho = 6.01 \text{ g cm}^{-3} \quad M_{AgI} = 234.77 \text{ g mol}^{-1} \quad S = 0.145 \text{ cm}^2\right.$

Exercise 2.6 – Determination of cationic transport number in CaF$_2$ by dilatocoulometry

The cationic transport number t_c in an M_aX_b ionic crystal is given by the following relation:
$$t_c = zF \times \frac{\rho}{M} \times \frac{S\Delta x}{I\tau}$$

where z is the cation charge number, F is the Faraday constant, ρ is the density of M_aX_b, M is the molar mass of M_aX_b, S is the surface area of the anode, Δx is the displacement of the anode-M_aX_b interface, I is the dc current crossing the anode, and τ is the electrolysis time.

The experimental setup is shown schematically in figure 25.

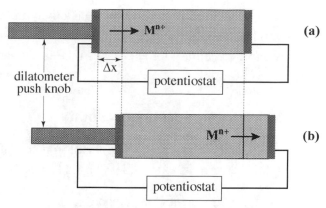

Figure 25 – Schematic of setup for dilatocoulometry
(a) before electrolysis, (b) after electrolysis.

We want to evaluate the cationic transport number of calcium fluoride, which crystallizes in a fluorine structure.

1. What precautions are needed in terms of gaseous atmosphere to measure the cationic transport number in CaF_2 by dilatocoulometry?

2. The cationic transport number obtained at 798 °C is 6.7×10^{-6} with two electrolysis currents: $I_1 = 0.15$ mA and $I_2 = 2$ mA.
   Comment on this result.

3. Table 17 gives the evolution of the total conductivity σ_t and of the transport number for CaF_2 for the extrinsic domain and as a function of temperature.

Table 17 – Variation of total conductivity and cationic transport number of CaF_2 as a function of temperature.

T [°C]	513	632	718	798	864
σ_t [S cm^{-1}]	8.66×10^{-7}	2.62×10^{-6}	4.86×10^{-6}	7.88×10^{-6}	1.11×10^{-5}
$t_c \times 10^6$	–	0.63	2	6.7	13

In what follows, we assume that the electronic conductivity in CaF_2 is negligible.

a. Draw the curve for total conductivity as a function of temperature in the form $\log(\sigma_t T) = f\left(\dfrac{10^3}{T}\right)$

b. Draw the curve for cationic conductivity as a function of temperature in the form $\log(\sigma_c T) = f\left(\dfrac{10^3}{T}\right)$

c. Deduce the activation energies for the total conductivity and the cationic conductivity. What can we conclude?

d. Give two examples of binary ionic crystals that have characteristics similar to those of CaF_2.

Exercise 2.7 – Electrochemical semipermeability

Consider a MX ionic crystal with the following chemical potential difference for the metal M applied between its extremities (1) and (2): $\mu_M^{(1)} - \mu_M^{(2)}$.

1. Give the expression for the potential difference
   a. ΔE_{1-2}^a for the case of a purely cationic conductor,
   b. ΔE_{1-2}^b for the case of a mixed conductor and as a function of the average cationic transport number $\overline{t_{M^+}}$.

2. In the absence of an external electric current, derive the expression for the gradient of the electrical potential, $\vec{\nabla} \varphi$, at an arbitrary point of a mixed conductor and as a function of the transport number t_{M^+}, of the gradients of the chemical potential, $\vec{\nabla} \mu_{M^+}$, of the cations M^+, and of the electrons ($\vec{\nabla} \mu_e$).

3. Deduce the expression of the particulate cationic current $\overrightarrow{I_{M^+}}$ as a function of the electronic conductivity σ_e, t_{M^+}, and $\vec{\nabla} \mu_M$.

4. Given that $\overrightarrow{I_{M^+}}$ is conserved, give the expression for the flux of species M^+ as a function of ΔE_{1-2}^a, of the total conductivity σ_t, of the average value $\overline{t_{M^+}(1 - t_{M^+})}$, and of the length ℓ of the sample.

Exercise 2.8 – Determination of transport number by electrochemical semipermeability

We want to determine the electronic transport number of a mixed-conductor solid solution (ionic by O^{2-} ions and p-type electronic). When a pellet of this compound separates two compartments T and CE containing gases with different oxygen partial pressures, the pellet is traversed by an oxygen flux called the specific flux J_{S,O_2}, which is expressed as

$$J_{S,O_2} = \frac{D}{V_m} \times \frac{\Delta P}{P_t} \times \frac{\ell}{S} \tag{1}$$

where D is the gas flow in compartment T, V_m is the molar volume, ΔP is the increase in oxygen partial pressure of compartment T (in bar), P_t is the total pressure in compartment T (in bar), ℓ is the thickness of the pellet, and S is the surface area of the pellet.

1. Using cm as units of length, give units for the flux J_{S,O_2}.

2. Calculate the specific flux of oxygen for the following conditions:
   Data

$T = 550\ °C$	$D = 4.25 \times 10^{-3}\ L\,s^{-1}$	$\Delta P = 1.932\ Pa$	$\ell = 0.2\ cm$
$V_m = 22.4\ L\,mol^{-1}$	$P_t = 1\ bar$	$S = 1.13\ cm^2$	

3. In the steady state, the oxygen partial pressure at the entrance of compartment T is 5×10^{-2} Pa and that of compartment CE is 0.21 bar (air). Calculate the potential difference ΔE between the two compartments when the pressure increase ΔP for compartment T is 1.932 Pa at 550 °C.

4. The total electric resistance R_t of the solid solution was measured in air at 550 °C. The geometric factor k of the sample is 0.755 cm^{-1}. The measured resistance increases to 6.29 Ω. Calculate the total conductivity of the solid solution.

5. Deduce the electronic transport number in air. What can we conclude about the validity of the answer to question 3?

Exercise 2.9 – Determination of conduction mode in α-AgI by Tubandt method

To determine the mode of conduction in silver iodide α-AgI, we use Tubandt method with the electrochemical setup shown schematically in figure 26.

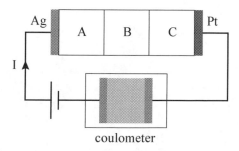

coulometer

Figure 26 – Setup to measure transport number of α-AgI.

Three separable α-AgI pellets are inserted between a silver anode and a platinum cathode. To measure the current, a silver coulometer is inserted in the circuit. The electrolyte of the coulometer consists of a solution of silver nitrate. The circuit carries a current of 100 mA for 1 h. The analysis of the results of this experiment leads to the following conclusions:

▷ no change is detected in the mass of tablets A and B,

▷ the mass of the silver anode decreases by Δm, which equals the mass deposited in the coulometer,

▷ the mass of tablet C remains constant,

▷ the mass of the platinum cathode increased by Δm.

1. What conclusions about the mode of conduction in α-AgI can we draw from this experiment?

2. Deduce the value of Δm.
   Data
   $$\left| \; M_{Ag} = 107.9 \text{ g mol}^{-1} \qquad M_I = 126.9 \text{ g mol}^{-1} \right.$$

Solutions to exercises

Solution 2.1 – Determination of conductivity by four-electrode method

1. Figure 27 shows the potential difference $V_B - V_C$ as a function of the current I passing through the cell. The curve is linear and is given by

$$V_B - V_C = 0.0422 \times I$$

with $V_B - V_C$ in mV and I in mA.

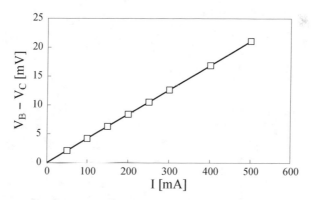

Figure 27 – Potential difference $V_B - V_C$ as a function of current I.

This linear variation indicates that $La_{0.8}Sr_{0.2}MnO_{3-\delta}$ is ohmic under the given measurement conditions.

2. The resistance R_1 between points B and C corresponds to the slope of the line and is given by $R_1 = 0.0422 \ \Omega$

3. The conductivity of $La_{0.8}Sr_{0.2}MnO_{3-\delta}$ is given by the relation

$$\sigma = \frac{k}{R} = \frac{7.59}{0.0422}$$

$$\sigma = 179.9 \ S \ cm^{-1}$$

4. **a.** Table 18 gives the data required to plot the conductivity in the form $\log \sigma T = f(1/T)$.

Table 18 – Log σT as a function of 1/T.

T [°C]	102	200	298	398	490	589	666	741
R [Ω]	0.1082	0.06938	0.05614	0.04954	0.04578	0.04357	0.04219	0.04125
T [K]	375	473	571	671	763	862	939	1014
$10^3/T$ [K^{-1}]	2.67	2.11	1.75	1.49	1.31	1.16	1.07	0.946
σ [S cm^{-1}]	70.15	109.4	135.2	153.2	165.8	174.2	180.3	184.0
σT [S cm^{-1} K]	26306	51746	77199	102797	126505	150160	169302	186576
log σT	4.42	4.714	4.888	5.012	5.102	5.176	5.23	5.271

The curve representing this function is linear (fig. 28). The expression for conductivity has the form

$$\log \sigma T = \log \sigma_0 - \frac{E_a}{2.3RT}$$

where E_a is the activation energy.

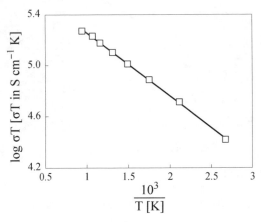

Figure 28 – Conductivity of $La_{0.8}Sr_{0.2}MnO_{3-\delta}$ plotted in the form log σT = f(1/T).

A linear regression gives the following relationship:

$$\log \sigma T = 5.7518 - \frac{0.4959}{T}$$

b. The activation energy E_a may be obtained either
   ▷ from the relation

$$E_a = 0.4971 \times 10^3 \times 8.314 \times \ln 10$$

$$E_a \approx 9506 \, J \, mol^{-1}$$

$$E_a \approx 0.1 \, eV$$

▷ or by choosing two pairs (σ, T), which leads to the following expression for the activation energy:

$$E_a = \frac{RT_1 T_2}{T_1 - T_2} \ln \frac{\sigma_1 T_1}{\sigma_2 T_2}$$

The pairs and lead to

$(T_1 = 375 \text{ K}, \sigma_1 = 70.15 \text{ S cm}^{-1})$
$(T_2 = 671 \text{ K}, \sigma_1 = 153.2 \text{ S cm}^{-1})$
$E_a = 9631 \text{ J mol}^{-1}$

$$E_a = 0.1 \text{ eV}$$

Note – This result is low compared with those associated with the classic ionic conductors (NaCl, CaF_2, YSZ).

5. **a.** The resistance measured between points A and B include the polarization resistance R_p of the electrode and the ohmic resistance R_2 of the material itself
$$R_{A\text{-}B} = R_p + R_2$$
If we assume a constant cross section for the sample, R_2 may be calculated from the following relation:

$$\frac{R_2}{R_1} = \frac{\ell_2}{\ell_1}$$

which gives

$$R_2 = R_1 \frac{\ell_2}{\ell_1}$$

$$R_2 = 0.0421 \times \frac{0.5}{1.496}$$

$$R_2 = 0.0141 \, \Omega$$

This result shows that the polarization resistance R_p is negligible ($R_p \approx 0$).

b. Taken together, the preceding results (low activation energy, negligible polarization resistance, role of the electrode material) indicate that $La_{0.8}Sr_{0.2}MnO_{3-\delta}$ is primarily an electronic conductor in air.

Solution 2.2 – Measurement of electric quantities by complex impedance spectroscopy

1. The equation for dipole devices associated in parallel is

$$\frac{1}{Z_{tot}} = \sum_k \frac{1}{Z_k}$$

which gives for two dipole devices 1 and 2 $Z_{tot} = \dfrac{Z_1 Z_2}{Z_1 + Z_2}$.

For the R//C circuit $\qquad Z_{R//C} = \dfrac{R}{1 + jRC\omega}$

The expression for the impedance of a R//C circuit in the form a + jb is

$$Z_{R//C} = \dfrac{R}{1 + R^2 C^2 \omega^2} - j\dfrac{R^2 C \omega}{1 + R^2 C^2 \omega^2}$$

This is the equation of a circle centered on the abscissa and passing through the origin.

For the circuit R//CPE, the equation for associating dipole devices gives

$$Z_{R//CPE} = \dfrac{R}{1 + RA(j\omega)^p}$$

Because $\qquad j^p = \cos\dfrac{p\pi}{2} + j\sin\dfrac{p\pi}{2}$

we obtain $\quad Z_{R//CPE} = \dfrac{R}{1 + RA\omega^p \cos\dfrac{p\pi}{2} + jRA\omega^p \sin\dfrac{p\pi}{2}}$

The expression for the impedance of a R//CPE circuit in the form a + jb is

$$Z_{R//CPE} = \dfrac{R\left(1 + RA\omega^p \cos\dfrac{p\pi}{2}\right)}{\left(1 + RA\omega^p \cos\dfrac{p\pi}{2}\right)^2 + \left(RA\omega^p \sin\dfrac{p\pi}{2}\right)^2}$$

$$- j\dfrac{R\left(RA\omega^p \sin\dfrac{p\pi}{2}\right)}{\left(1 + RA\omega^p \cos\dfrac{p\pi}{2}\right)^2 + \left(RA\omega^p \sin\dfrac{p\pi}{2}\right)^2}$$

This is the equation for a circle with its center under the abscissa and that passes through the origin. For p = 1, we find the expression for the impedance of the a R//C circuit (ideal capacitor, zero decentering).

2. The complex impedance of a R//CPE circuit is represented in the complex plane in figure 29. β_1 and β_2 are the decentering angles for p = 0.75 and p = 0.5, respectively.

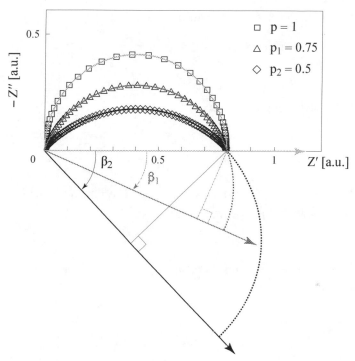

Figure 29 – Complex impedance spectra for R//CPE circuit for various values of p.

3. Here we present the calculations for the R//CPE circuit. The calculations for a R//C circuit are analogous.

   a. The geometric factor is calculated as follows:

   $$k = \frac{\ell}{S} = \frac{0.177}{1.267}$$

   $$k = 0.14\,\text{cm}^{-1} = 14\,\text{m}^{-1}$$

   b. The relaxation frequency is given by

   $$\omega_0 = (RA)^{-\frac{1}{p}} = (6.19 \times 10^4)(1.41 \times 10^{-10})^{-\frac{1}{0.879}}$$

   $$\omega_0 = 5.69 \times 10^5 \,\text{rad s}^{-1}$$

   c. The time constant (relaxation time) is the inverse of ω_0 within 2π

   $$\tau_0 = \frac{2\pi}{\omega_0} = \frac{2\pi}{5.69 \times 10^5}$$

   $$\tau_0 = 1.1 \times 10^5 \,\text{s}$$

d. The capacitance C of the sample is

$$C = \frac{1}{R\,\omega_0} = \frac{1}{(6.19 \times 10^4)(5.69 \times 10^5)}$$

$$C = 2.84 \times 10^{-11} \text{ F}$$

e. The decentering angle β is

$$\beta = (1-p)\frac{\pi}{2} = (1-0.879)\frac{\pi}{2}$$

$$\beta = 10.9°$$

f. The capacitance C of the sample depends on the dielectric constant ε_r of the sample and its geometric factor k:

$$C = \varepsilon_0\varepsilon_r \times \frac{S}{\ell} = \frac{\varepsilon_0\varepsilon_r}{k}$$

where ε_0 is the vacuum permittivity with $\varepsilon_0 = 10^7/(4\pi c^2)$ and c is the speed of light.

$$\varepsilon_0 = \frac{10^7}{4\pi \times (3 \times 10^8)^2}$$

$$\varepsilon_0 = 8.85 \times 10^{-12} \text{ F m}^{-1}$$

From this we deduce

$$\varepsilon_r = \frac{C\,k}{\varepsilon_0} = \frac{2.84 \times 10^{-11} \times 14}{8.85 \times 10^{-12}}$$

$$\varepsilon_r = 44.9$$

g. The permittivity is thus

$$\varepsilon = 44.9 \times 8.85 \times 10^{-12}$$

$$\varepsilon = 3.97 \times 10^{-10} \text{ F m}^{-1}$$

h. The conductivity of the sample is given by

$$\sigma = \frac{1}{R} \times \frac{\ell}{S} = \frac{k}{R}$$

or

$$\sigma = \frac{0.14}{6.19 \times 10^4}$$

$$\sigma = 2.26 \times 10^{-6} \text{S cm}^{-1}$$

i. The expression relating resistivity ρ to the conductivity σ is

$$\rho = \frac{1}{\sigma}$$

$$\rho = 4.42 \times 10^5 \,\Omega\,\text{cm}$$

4. Resistivity, conductivity, dielectric constant, permittivity, and relaxation frequency are all characteristics that are theoretically independent of sample geometry.

5. Given the results of the analysis of residues and the agreement between the calculated points and the observations (see fig. 22), the most appropriate circuit is the R//CPE circuit.

Solution 2.3 – Measurement of electronic conductivity in a mixed conductor

1. The scheme for the electrochemical chain is

$$O_2, Pt/CuO\text{-}Cu_2O/(MIEC)/Pt, O_2$$

2. At the electrodes, the reactions are

$$\tfrac{1}{2}O_{2(g)}^{WE} + 2e_{Pt} \rightleftarrows O_{MIEC}^{2-}$$

3. The equilibrium conditions of the system at the reference electrode are

$$\tfrac{1}{2}\mu_{O_2}^{RE} + 2\tilde{\mu}_e^{RE} = \tilde{\mu}_{O^{2-}}^{RE}$$

and at the working electrode, $\tfrac{1}{2}\mu_{O_2}^{WE} + 2\tilde{\mu}_e^{WE} = \tilde{\mu}_{O^{2-}}^{WE}$

The equilibrium conditions between the various phases allow us to write

$$\tilde{\mu}_{O^{2-}}^{WE} = \tilde{\mu}_{O^{2-}}^{RE}$$

because $J(O^{2-}) = 0$ (blocking electrode),

with
$$4F(\varphi^{WE} - \varphi^{RE}) = \left(\mu_{O_2}^{WE} - \mu_{O_2}^{RE}\right)$$

and
$$U = \varphi^{RE} - \varphi^{WE} = \frac{1}{4F}\left(\mu_{O_2}^{WE} - \mu_{O_2}^{RE}\right) = \frac{1}{4F}\Delta\mu_{O_2}$$

where φ^i is the electrical potential of phase i.

Recalling that
$$\mu_{O_2}^i = \mu_{O_2}^{0,i} + RT \ln a_{O_2}^i$$

we finally obtain
$$U = \frac{RT}{4F} \ln \frac{a_{O_2}^{WE}}{a_{O_2}^{RE}}$$

from which we deduce
$$a_{O_2}^{WE} = a_{O_2}^{RE} e^{\frac{4UF}{RT}}$$

4. **a.** At 750 °C, the oxygen activity at the working electrode is $a_{O_2}^{WE} = a_{O_2}^{RE} e^{\frac{4UF}{RT}}$

with $\qquad a_{O_2}^{RE}(750\,°C) = 2.68 \times 10^{-4}$

For U = −0.5 V $a_{O_2}^{WE} = 268 \times 10^{-4}\, e^{\frac{4 \times (-0.5) \times 96480}{8.314 \times (750+273)}}$

$$a_{O_2}^{WE} = 3.76 \times 10^{-14}$$

and $\qquad\qquad\qquad \log a_{O_2}^{WE} = -13.42$

The results for $\log a_{O_2}^{WE}$ are presented in table 19.

Table 19 – Oxygen activity at working electrode as a function of applied potential.

U [V]	− 0.5	− 0.4	− 0.3	− 0.2	− 0.1	0	0.1	0.2	0.3
$\log a_{O_2}^{WE}$	− 13.42	− 11.45	− 9.48	− 7.51	− 5.54	− 3.57	− 1.60	0.37	2.34

b. The curves for $I(U)$ and $\log a_{O_2}^{WE}(U)$ are shown in figures 30 and 31, respectively.

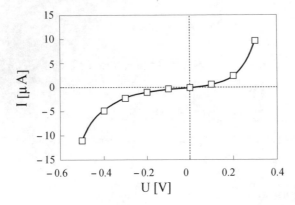

Figure 30 – Current plotted as a function of applied potential.

As shown in figure 31, the curve $\log a_{O_2}^{WE}(U)$ is a line. A linear regression gives $\qquad \log a_{O_2}^{WE} = 19.70\, U - 3.57$

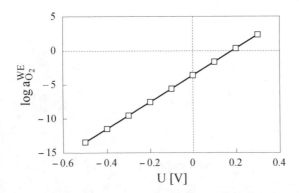

Figure 31 – Plot of log of oxygen activity at working electrode as a function of applied potential.

c. Between the points U = −0.5 V and U = −0.4 V, which gives an average applied potential of −0.45 V, the equation of the line in figure 31 allows us to calculate $\log a_{O_2}^{WE}$.

$$\log a_{O_2}^{WE} = 19.70 \times \frac{-0.4 - 0.5}{2} - 3.57$$

$$\log a_{O_2}^{WE} = -12.44$$

Furthermore, the electronic conductivity is calculated by applying the relation

$$\sigma_e(a_{O_2}^{WE}) = \frac{1}{2\pi a} \times \frac{dI}{dU}$$

and averaging two successive values of current and potential to calculate the derivative. For example, by using the two pairs (−4.8/−0.4) and (−11/−0.5), we obtain

$$\sigma_e(a_{O_2}^{WE}) = \frac{1}{2\pi \times 170 \times 10^{-4}} \times \frac{(-4.8 + 11) \times 10^{-6}}{(-0.4 + 0.5)}$$

$$\sigma_e(a_{O_2}^{WE}) = 5.80 \times 10^{-4}\,S\,cm^{-1}$$

or

$$\log \sigma_e(a_{O_2}^{WE}) = -3.24\,S\,cm^{-1}$$

The results for $\log a_{O_2}^{WE}$ and $\log \sigma_e(a_{O_2}^{WE})$ for each average potential U_{ave} appear in table 20.

Table 20 – Logarithm of oxygen activity at working electrode and of electronic conductivity as a function of average applied potential.

U_{ave} [V]	−0.45	−0.35	−0.25	−0.15	−0.05	0.05	0.15	0.25
$\log a_{O_2}^{WE}$	−12.44	−10.47	−8.50	−6.53	−4.56	−2.59	−0.62	1.36
$\log \sigma_e$ [$S\,cm^{-1}$]	−3.24	−3.61	−3.95	−4.25	−4.43	−4.25	−3.77	−3.17

5. Figure 32 shows the CGP electronic conductivity as a function of oxygen activity on a logarithmic scale.

The branches σ_n and σ_p are identified based on the slope of the lines. A linear regression through the points of each branch gives

  ▷ for branch n $\log \sigma_n = -0.18 \times \log a_{O_2}^{WE} - 5.49$

  ▷ for branch p $\log \sigma_p = 0.27 \times \log a_{O_2}^{WE} - 3.56$

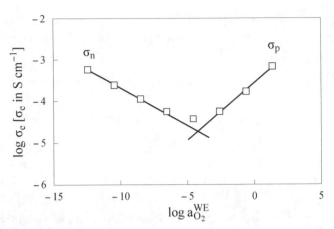

Figure 32 – Log of oxygen activity as a function of applied potential.

6. The equations for the lines obtained in question 5 allow us to calculate the conductivities σ_n and σ_p for $a_{O_2}^{WE} = 1$ and $a_{O_2}^{WE} = 10^{-12}$.

For $a_{O_2}^{WE} = 10^{-12}$ $\log \sigma_n = -0.18 \times \log(10^{-12}) - 5.49$

$$\log \sigma_n = -3.32$$

or $\sigma_n = 4.68 \times 10^{-4}\,S\,cm^{-1}$

For $a_{O_2}^{WE} = 1$ $\log \sigma_p = 0.27 \times \log(1) - 3.56$

$$\log \sigma_p = -3.56$$

or $\sigma_p = 2.75 \times 10^{-4}\,S\,cm^{-1}$

Solution 2.4 – Measurement of ionic conductivity in a mixed conductor

1. The scheme for the electrochemical chain is

$$O_2, Pt\,/\,YSZ\,/\,LSCo, O_2$$

2. The two electrodes (WE and CE) are subject to the following equilibrium:

$$\tfrac{1}{2}O_{2(g)} + 2e \rightleftharpoons O^{2-}$$

which allows us to write

▷ at the counter electrode

$$\tfrac{1}{2}\mu_{O_2}^{CE} + 2\tilde{\mu}_e^{CE} = \tilde{\mu}_{O^{2-}}^{CE}$$

▷ at the working electrode

$$\tfrac{1}{2}\mu_{O_2}^{WE} + 2\tilde{\mu}_e^{WE} = \tilde{\mu}_{O^{2-}}^{WE}$$

The equilibrium conditions between the various phases leads to

$$\tilde{\mu}_{O^{2-}}^{WE} = \tilde{\mu}_{O^{2-}}^{CE}$$

because $J(e) = 0$ (blocking electrode),

with

$$4F(\varphi^{CE} - \varphi^{WE}) = \mu_{O_2}^{WE} - \mu_{O_2}^{CE}$$

and

$$U = \varphi^{CE} - \varphi^{WE} = \frac{1}{4F}\left(\mu_{O_2}^{WE} - \mu_{O_2}^{CE}\right) = \frac{1}{4F}\Delta\mu_{O_2}$$

where φ^i is the electrical potential of phase i.

Given that

$$\mu_{O_2}^i = \mu_{O_2}^{0,i} + RT \ln a_{O_2}^i$$

we finally obtain

$$U = \frac{RT}{4F} \ln \frac{a_{O_2}^{WE}}{a_{O_2}^{CE}}$$

or

$$a_{O_2}^{WE} = a_{O_2}^{CE} e^{\frac{4UF}{RT}}$$

3. The curve I(U) is given in figure 33.

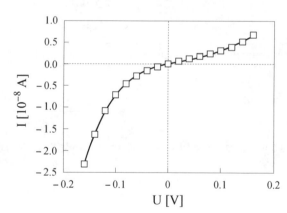

Figure 33 – Current as a function of applied potential.

A polynomial regression leads to an analytic expression for the current as a function of applied potential. This expression is a fourth-order polynomial:

$$I = -7 \times 10^{-6} \times U^4 + 3 \times 10^{-6} \times U^3 - 10^{-7} \times U^2 + 3 \times 10^{-8} \times U + 5 \times 10^{-11}$$

with U in V and I in A.

The oxygen activity at the working electrode is derived from the expression obtained in question 2. The oxygen activity at the counter electrode corresponds to the oxygen partial pressure in air

$$a_{O_2}^{CE} = 0.2$$

The ionic conductivity of the sample is calculated from the expression

$$\sigma_i(a_{O_2}^{WE}) = \frac{1}{2\pi a} \times \frac{dI}{dU}$$

The factor dI/dU is obtained by calculating the derivative of the analytical expression for current as a function of the applied potential determined above:

$$\frac{dI}{dU} = -2.8 \times 10^{-5} \times U^3 + 9 \times 10^{-6} \times U^2 - 2 \times 10^{-7} \times U + 3 \times 10^{-8}$$

with U in V and I in A.

The results for the oxygen activity $a_{O_2}^{WE}$ and the ionic conductivity σ_i of the sample for each value of the applied potential appear in table 21.

Table 21 – Logarithm of oxygen activity and of the ionic conductivity of LSCo as a function of applied potential U.

U [V]	− 0.160	− 0.140	− 0.120	− 0.100	− 0.080	− 0.060	− 0.040	− 0.020
$\log a_{O_2}^{WE}$	− 4.39	− 3.93	− 3.47	− 3.01	− 2.55	− 2.08	− 1.62	− 1.16
$\log \sigma_i$ [S cm^{-1}]	− 5.19	− 5.30	− 5.43	− 5.57	− 5.73	− 5.89	− 6.06	− 6.22

U [V]	0	0.020	0.040	0.060	0.080	0.100	0.120	0.140	0.160
$\log a_{O_2}^{WE}$	− 0.70	− 0.24	0.22	0.69	1.15	1.61	2.07	2.53	3.00
$\log \sigma_i$ [S cm^{-1}]	− 6.32	− 6.33	− 6.26	− 6.15	− 6.04	− 5.94	− 5.86	− 5.79	− 5.74

4. Figure 34 shows the ionic conductivity of LSCo as a function of oxygen activity.

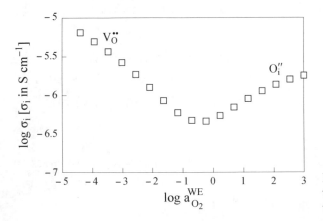

Figure 34 – Ionic conductivity of LSCo as a function of oxygen activity.

Two phenomena may be observed:

▷ The conductivity increases when the oxygen activity decreases. This behavior may be explained by the fact that the ionic mobility is based on vacancies. When $a_{O_2}^{WE}$ decreases, the proportion of oxygen vacancies, which are the charge carriers, increases, thereby increasing the conductivity.

▷ For strong oxygen activity, the conductivity increases with a more gentle slope, which may be explained by the eventual formation of interstitial oxygen.

Solution 2.5 – Determination of cationic transport number by dilatocoulometry

1. Figure 35 shows the principle of dilatocoulometry.

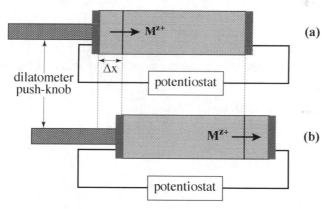

Figure 35 – Schematic diagram illustrating the displacement of the interphases Me,MX upon passage of current **(a)** before electrolysis and **(b)** after electrolysis.

Consider the following electrochemical chain:

$$X_2, Me / MX / Me, X_2$$

where X_2 is the molecular form for the gaseous element X, Me is an inert metallic electrode, and MX is the ionic compound under study.

When an electric current I runs through this chain, matter is transported because of a partial cationic conduction in MX. This transport leads to a displacement of the interphases Δx, which is followed by dilatometry.

2. The displacement of the interface at the anode is due to the disappearance of the quantity n_+ of substance in the form of the species MX, which in turn requires the species M^{z+} to transport the following quantity of charge:

$$Q_+ = t_c I \tau$$

Q_+ may also be expressed as $Q_+ = n_+ z F$

We arrive at
$$t_c = z F \frac{n_+}{I \tau}$$

n_+ (or n_{MX}) can be expressed as a function of the density ρ and the molar mass M of compound MX

$$n_{MX} = \frac{m_{MX}}{M} = \frac{\rho \times S \times \Delta x}{M}$$

The final expression for the cationic transport number M^{z+} in MX is

$$t_c = z F \times \frac{\rho}{M} \times \frac{S \times \Delta x}{I \times \tau}$$

3. **a.** To liberate gaseous iodine, we must ensure that the temperature used exceeds the sublimation temperature of iodine I_2 under atmospheric pressure.

   b. The following relation should be applied:
$$t_c = F \times \frac{\rho}{M} \times \frac{S \times \Delta x}{I \times \tau}$$

   by using $I \times \tau = Q$, we obtain
$$t_c = 358.13 \times \frac{\Delta x}{Q}$$

   Δx is expressed in cm and Q in coulombs.
   Applying this to the two temperatures under consideration gives

T [°C]	250	300
t_c	0.977	1.005

   c. The results confirm that, in α-AgI, the cationic transport number is practically equal to unity.

Solution 2.6 – Determination of cationic transport number in CaF$_2$ by dilatocoulometry

1. We must ensure that the oxygen partial pressure is sufficiently low to minimize the introduction of oxygen into the crystal. Under these conditions, the

anodic reaction corresponds to a release of fluorine F_2 at the total working pressure:

$$2\,F_F^\times \longrightarrow 2\,V_F^\bullet + F_{2(g)} + 2e'$$

2. This behavior indicates that the variation in current I does not lead to a modification of the crystal and, in particular, of the anodic reaction. Applying the relation for t_c, to CaF_2 gives

$$t_c = 2F \times \frac{\rho}{M} \times \frac{S \times \Delta x}{I \times \tau}$$

This shows that, for a given electrolysis time τ, any variation in I is accompanied by a variation in the ratio $\Delta x / \tau$ such that the ratio $\Delta x / (I \times \tau)$ and, consequently, t_c, remains constant.

3. **a.** The curve corresponding to the function $\log(\sigma_1 T) = f(10^3/T)$ is linear, as shown in figure 36. This indicates that the total conductivity is an activated process (see chap. 3).

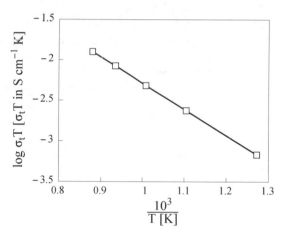

Figure 36 – Total conductivity of CaF_2 in Arrhenius coordinates.

b. The cationic transport number is defined as the ratio of the cationic conductivity to the total conductivity, $t_c = \sigma_c / \sigma_t$.
From this we deduce the cationic conductivity as

$$\sigma_c = \sigma_t \times t_c$$

The cationic conductivity of CaF_2 is given in table 22 as a function of temperature.

Table 22 – Cationic conductivity of CaF_2 as a function of temperature.

T [°C]	632	718	798	864
σ_c [S cm^{-1}]	1.65×10^{-12}	9.72×10^{-12}	5.28×10^{-11}	1.44×10^{-10}

The curve corresponding to the function $\log(\sigma_c T) = f(10^3/T)$ is also linear, as shown in figure 37. This indicates that the cationic conduction is an activated process in the temperature range considered.

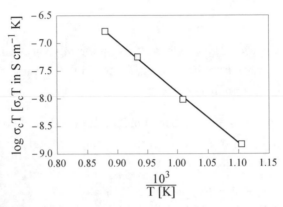

Figure 37 – Cationic conductivity of CaF_2 in Arrhenius coordinates.

Remark – The results obtained for the cationic conductivity show that it is negligible with respect to the total conductivity. For example, the ratio σ_c/σ_t is 2×10^{-6} at 718 °C. We can thus conclude that anionic conductivity dominates in CaF_2.

c. A linear regression based on the experimentally determined data yields the lines shown in figures 36 and 37.

This gives for the total conductivity

$$\log(\sigma_t T) = -3.228 \times \frac{10^3}{T} + 0.940$$

and for the cationic conductivity

$$\log(\sigma_c T) = -9.154 \times \frac{10^3}{T} + 1.269$$

The Arrhenius equation is $\sigma = \dfrac{\sigma_0}{T} e^{\frac{-E_a}{RT}}$

where E_a is the activation energy for ionic migration. We can deduce the corresponding activation energy

▷ for total conductivity

$$E_{a,t} = 0.64 \text{ eV}$$

▷ for cationic conductivity

$$E_{a,c} = 1.81\,eV$$

We observe a significant difference between the two activation energies, which is shown by the difference between the anionic and cationic conductivities.

d. We find behavior similar to that of CaF_2 in terms of ionic-conducting substance (i.e., a dominant anionic conduction) such as ionic crystals with the same structure (e.g., BaF_2, $SrCl_2$, PbF_2, ThO_2, CeO_2, UO_2, ZrO_2, Bi_2O_3, …).

Solution 2.7 – Electrochemical semipermeability

1. **a.** We have a mixed conductor containing cations M^+ and electrons that are subjected to a difference in chemical potential between the two extremities of a metal M. The general expression for the potential difference ΔE between terminals is

$$\Delta E = \frac{1}{F}\int_1^2 t_{ion}\,d\mu_M$$

where t_{ion} is the ionic transport number of the crystal, which here equals the cationic transport number t_{M^+}.

For a purely cationic conductor ($t_{M^+} = 1$), we obtain

$$\Delta E_{1-2}^a = \frac{1}{F}\left(\mu_M^{(2)} - \mu_M^{(1)}\right)$$

b. If we consider the average ionic transport number $\overline{t_{M^+}}$ in the domain $\mu_M^{(1)} - \mu_M^{(2)}$, the preceding relation becomes

$$\Delta E = \frac{\overline{t_{M^+}}}{F}\int_1^2 d\mu_M$$

$$\Delta E = \frac{\overline{t_{M^+}}}{F}\left(\mu_M^{(2)} - \mu_M^{(1)}\right)$$

and the electromotive force ΔE_{1-2}^b is

$$\Delta E_{1-2}^b = \frac{\overline{t_{M^+}}}{F}\left(\mu_M^{(2)} - \mu_M^{(1)}\right) = \overline{t_{M^+}}\,\Delta E_{1-2}^a$$

2. The cationic and electric current densities are given by the following expressions

▷ cationic current $\quad \overrightarrow{i_{M^+}} = -\frac{\sigma_{M^+}}{F}\left(\overrightarrow{\nabla}\mu_{M^+} + F\overrightarrow{\nabla}\varphi\right)$

▷ electronic current $\vec{i_e} = \dfrac{\sigma_e}{F}(\vec{\nabla}\mu_e - F\vec{\nabla}\varphi)$

Given that the external current is zero, we can write

$$0 = \vec{i_{M^+}} + \vec{i_e}$$

$$0 = -\frac{\sigma_{M^+}}{F}(\vec{\nabla}\mu_{M^+} + F\vec{\nabla}\varphi) + \frac{\sigma_e}{F}(\vec{\nabla}\mu_e - F\vec{\nabla}\varphi)$$

and, by using $\sigma_t = \sigma_{M^+} + \sigma_e$ for the total conductivity, we obtain

$$0 = -\sigma_t\vec{\nabla}\varphi - \frac{\sigma_{M^+}}{F}\vec{\nabla}\mu_{M^+} + \frac{\sigma_e}{F}\vec{\nabla}\mu_e$$

Given that $\quad t_{M^+} = \dfrac{\sigma_{M^+}}{\sigma_t}\quad$ and $\quad t_e = \dfrac{\sigma_e}{\sigma_t} = 1 - t_{M^+}$

we derive $\qquad \vec{\nabla}\varphi = -\dfrac{t_{M^+}}{F}\vec{\nabla}\mu_{M^+} + \dfrac{(1 - t_{M^+})}{F}\vec{\nabla}\mu_e$

3. The equality $i_{M^+} = -i_e$ allows us to write

$$-\frac{\vec{i_{M^+}}}{\sigma} = \vec{\nabla}\mu_e - F\vec{\nabla}\varphi$$

By replacing $\vec{\nabla}\varphi$ by the expression established in question 2, we have

$$-\frac{\vec{i_{M^+}}}{\sigma_e}F = \vec{\nabla}\mu_e + t_{M^+}\vec{\nabla}\mu_{M^+} - (1 - t_{M^+})\vec{\nabla}\mu_e$$

and $\qquad -\dfrac{\vec{i_{M^+}}}{\sigma_e}F = t_{M^+}(\vec{\nabla}\mu_{M^+} + \vec{\nabla}\mu_e)$

The equilibrium $\qquad M \rightleftharpoons M^+ + e$

leads to $\qquad \vec{\nabla}\mu_{M^+} + \vec{\nabla}\mu_e = \vec{\nabla}\mu_M$

which is to say $\qquad -\dfrac{\vec{i_{M^+}}}{\sigma_e}F = -t_{M^+}\vec{\nabla}\mu_M$

or $\qquad \vec{i_{M^+}} = -t_{M^+}\dfrac{\sigma_e}{F}\vec{\nabla}\mu_M$

4. The expression relating the flux to the ionic current is

$$\vec{J_{M^+}} = \frac{\vec{i_{M^+}}}{F} = \frac{\sigma_e t_{M^+}}{F^2}\vec{\nabla}\mu_M$$

$$J_{M^+}d\ell = \frac{\sigma_e t_{M^+}}{F^2}d\mu_M$$

Because the ionic current is conservative, we can write

$$\int_1^2 J_{M^+}d\ell = |J_{M^+}| \times \ell = \int_1^2 \frac{\sigma_e t_{M^+}}{F^2}d\mu_M$$

Taking the average values, we obtain

$$|J_{M^+}| \times \ell = \frac{\sigma_t}{F^2}\left(1 - t_{M^+}\right)t_{M^+}\left(\mu_M^{(2)} - \mu_M^{(1)}\right)$$

Because $\Delta E_{1-2}^a = \frac{1}{F}\left(\mu_M^{(2)} - \mu_M^{(1)}\right)$, we obtain

$$|J_{M^+}| = \frac{\sigma_t(1 - t_{M^+})t_{M^+}}{F\ell}\Delta E_{1-2}^a$$

Solution 2.8 – Determination of transport number by electrochemical semipermeability

1. The unit of the flux J_{S,O_2}

   The dimensional equation applied to equation (1) gives

   $$J_{S,O_2} \equiv \frac{cm^2\,s^{-1}\,cm}{cm^3\,mol^{-1}\,cm^2}$$

   or $J_{S,O_2} \equiv mol\,s^{-1}\,cm^{-2}$

2. Numerically evaluating equation (1) gives

   $$J_{S,O_2} = \frac{4.25 \times 10^{-3}}{22.4} \times \frac{1.932}{10^5} \times \frac{0.2}{1.13}$$

   $$J_{S,O_2} = 6.49 \times 10^{-10}\,mol\,s^{-1}\,cm^{-1}$$

3. The expression for the potential difference is

   $$\Delta E = \frac{RT}{4F}\ln\frac{P'_T}{P_{CE}}$$

   with $P'_T = P_1 + \Delta P$

   $$P'_T = 5 \times 10^{-2} + 1.932$$

   $$P'_T = 1.982\ Pa$$

   and $\Delta E = \frac{RT}{4F}\ln\frac{P'_T}{P_{CE}}$

   $$\Delta E = \frac{8.314 \times 823}{4 \times 96\,480}\ln\frac{1.982}{0.21 \times 10^5}$$

   $$\Delta E = -0.164\ V$$

4. The total conductivity σ_t of the solid solution is obtained from the relation

   $$\sigma_t = \frac{k}{R_t}$$

Numerical evaluation gives $\sigma_t = \dfrac{0.755}{6.29}$

$$\sigma_t = 0.12\,\text{S cm}^{-1}$$

5. The expression relating the electronic conductivity to the specific oxygen flux measured by semipermeability is

$$\sigma_h = \frac{4\,F^2}{RT}\,J_{S,O_2}$$

$$\sigma_h = \frac{4\,F^2}{RT}\,J_{S,O_2} = \frac{4 \times 96\,480^2}{8.314 \times 823} \times 6.49 \times 10^{-10}$$

$$\sigma_h = 3.53 \times 10^{-3}\,\text{S cm}^{-1}$$

The electronic transport number t_h is

$$t_h = \frac{\sigma_h}{\sigma_t}$$

$$t_h = \frac{3.53 \times 10^{-3}}{0.12}$$

$$t_h = 2.9 \times 10^{-2}$$

The relatively high electronic transport number should decrease the potential difference calculated in question 3.

Solution 2.9 – Determination of conduction mode in α-AgI by Tubandt method

1. ▷ The first two observations allow us to conclude that the electronic transport number in α-AgI is negligible ($t_{el} \approx 0$),

   ▷ the constant mass of pellet C indicates that the anionic transport number in α-AgI is negligible ($t_{an} \approx 0$),

   ▷ the mass variations in the anode and the cathode indicate that α-AgI is a purely cationic conductor ($t_c \approx 1$).

2. The variation in mass is given by

$$\Delta m = \frac{Q}{F}M_{Ag} = \frac{It}{F}M_{Ag}$$

$$\Delta m = \frac{0.1 \times 3600}{96\,480} \times 107.9$$

$$\Delta m = 402.6\,\text{mg}$$

Chapter 3

Transport in ionic solids

Course notes

The displacement of ions in an ionic crystal can be described phenomenologically by following the approach developed by electrochemists, or microscopically by following the approach developed, in particular, by physicists.

3.1 – Phenomenological approach to ionic transport in ionic crystals

3.1.1 – Electrochemical mobility and flux density

Consider a chemical species i subject to a driving force Γ_m resulting from the action of an electrochemical potential gradient $\vec{\nabla}\tilde{\mu}_i$

$$\vec{F_m} = -\vec{\nabla}\tilde{\mu}_i \qquad \text{with } \tilde{\mu}_i = \mu_i + z_i F\varphi$$

where μ_i and z_i represent respectively the chemical potential and the charge number of species i, and φ is the electrical potential of the phase. Opposing this driving force is a friction force $|F_i| = \alpha v_i$ and an inertial force $|F_i| = m_i\,dv_i/dt$ where m_i and α denote respectively the particle mass and the friction coefficient. In the steady state, the forces sum to zero. We show that the particle rapidly attains a limiting speed $v_{i,t}$ given by the expression $\vec{v_{i,t}} = \vec{\nabla}\tilde{\mu}_i/\alpha$. We use the following definitions:

▷ the limiting speed per unit force is the electrochemical mobility

$$\tilde{u}_i = -\frac{\left|\vec{v_{i,t}}\right|}{\left|\vec{\nabla}\tilde{\mu}_i\right|} = \frac{1}{\alpha}$$

© Springer Nature Switzerland AG 2020
A. Hammou and S. Georges, *Solid-State Electrochemistry*,
https://doi.org/10.1007/978-3-030-39659-6_3

▷ the limiting speed per unit field is the electric mobility $u_i = -z_i F \dfrac{\left|\overrightarrow{v_{i,t}}\right|}{\left|\overrightarrow{\nabla}\tilde{\mu}_i\right|}$

The flux density J_i of particle i is given by the product of its speed and its concentration C_i, which gives:

$$\overrightarrow{J_i} = -C_i\tilde{u}_i\overrightarrow{\nabla}\mu_i - z_iFC_i\tilde{u}_i\overrightarrow{\nabla}\varphi$$

Application to pure chemical diffusion ($\overrightarrow{\nabla}\varphi = \overrightarrow{0}$) gives:

$$\overrightarrow{J_i} = -C_i\tilde{u}_i\overrightarrow{\nabla}\mu_i$$

Given that $\mu_i = \mu_i^\circ + RT\ln a_i$ and equating the activity a_i with the concentration C_i, we obtain:

$$\overrightarrow{J_i} = -RT\tilde{u}_i\overrightarrow{\nabla}C_i$$

By comparing this relation to Fick's first law $\overrightarrow{J_i} = -D_i\overrightarrow{\nabla}C_i$, where D_i is the diffusion coefficient of species i, we deduce that $D_i = RT\tilde{u}_i$.

3.1.2 – Electrical conductivity and transport number

◇ *Particulate electrical conductivity σ_i*

In the case of electric migration ($\overrightarrow{\nabla}\mu_i = 0$), the flux density is

$$\overrightarrow{J_i} = -z_iFC_i\tilde{u}_i\overrightarrow{\nabla}\varphi$$

from which we deduce the particulate current density i_i

$$\overrightarrow{i_i} = z_iF\overrightarrow{J_i} = -z_i^2F^2C_i\tilde{u}_i\overrightarrow{\nabla}\varphi$$

Identifying this with Ohm's law $i_i = \sigma_iE = -\sigma_i\left|\overrightarrow{\nabla}\varphi\right|$ gives us the expression for the particulate electrical conductivity σ_i

$$\sigma_i = z_i^2F^2C_i\tilde{u}_i = z_iFC_iu_i$$

◇ *Total electrical conductivity σ_t*

The total electrical conductivity σ_t is the sum of the particulate electrical conductivities

$$\sigma_t = \sum_i \sigma_i$$

taking into consideration all the ionic and electronic species that participate in transport.

◇ *Transport number t_i*

The transport number of species i is defined by the relation

$$t_i = \frac{\sigma_i}{\sigma_t}$$

The sum of the transport numbers must satisfy the following normalization condition:
$$\sum_i t_i = 1$$

◇ *Nernst-Einstein relation*

Considering that the diffusion and migration mechanisms are identical, we have

$$\sigma_i = \frac{z_i^2 F^2 C_i D_i}{RT}$$

3.2 – Microscopic approach to ionic transport in crystals: Activated-hopping model

3.2.1 – Electric mobility

The ion, which can move, oscillates in a potential well with a vibrational frequency ν_0 of the order of 10^{13} Hz. We allow the ion to hop from one site to a neighboring site along the x direction only and according to a unique mechanism. With no external electric field the forward and reverse hopping frequencies ν_f and ν_r are equal and are given by

$$\nu_f = \nu_r = \nu_0 e^{-\frac{\Delta_m G}{RT}}$$

where $\Delta_m G$ is the free enthalpy for the ion to attain the activated state and acts as a "potential barrier" (fig. 38). R and T have their usual meanings.

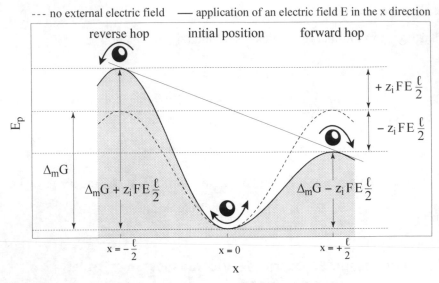

Figure 38 – Energy diagram for displacement of
an ion in an ionic crystal (from Déportes *et al.*, 1994).

When an electric field E is applied in the x direction, the expressions for the
hopping frequencies become

$$\nu_f = \nu_0 \, e^{-\frac{\Delta_m G - z_i FE \,\ell/2}{RT}} \quad \text{and} \quad \nu_r = \nu_0 \, e^{-\frac{\Delta_m G + z_i FE \,\ell/2}{RT}}$$

where z_i is the charge number of the ion and $\ell/2$ is the hopping distance. Under
these conditions, the effective hopping frequency is

$$\nu = \nu_f - \nu_i = \nu_0 \left(e^{\frac{z_i FE \,\ell/2}{RT}} - e^{-\frac{z_i FE \,\ell/2}{RT}} \right) e^{-\frac{\Delta_m G}{RT}}$$

Under conditions typically used for measurement and for exploiting the transport properties of ionic crystals, we have $\frac{z_i FE \,\ell/2}{RT} \ll 1$. After expanding the
exponential in a power series, this gives

$$\nu = \frac{z_i FE \,\ell}{RT} \nu_0 \, e^{-\frac{\Delta_m G}{RT}}$$

The distance x covered by the ion in time t is $x = \nu \ell t$, which gives an average
speed $\overline{v_i}$ of the ion of

$$\overline{v_i} = \frac{z_i FE \,\ell^2}{RT} \nu_0 \, e^{-\frac{\Delta_m G}{RT}}$$

By defining the electric mobility u_i of the ion as its average speed per unit
field, we have

$$u_i = \frac{z_i F \ell^2}{RT} \nu_0 \, e^{-\frac{\Delta_m G}{RT}}$$

Note – In reality, the ion hops are correlated. This results in the introduction of a correlation factor f in the expression for ionic migration that, as a simplification, is generally taken to be unity. In the framework of the activated hopping model, the ionic mobility is

$$u_i = f \frac{z_i F \ell^2}{RT} v_0 e^{-\frac{\Delta_m G}{RT}}$$

3.2.2 – Ionic conductivity

The thermodynamics of irreversible processes shows that the conductivity σ_i of ion i is given by the relation $\sigma_i = z_i F u_i N x_i (1 - x_i)$, where N is the site density (expressed in $mol\,m^{-3}$) and x_i is the molar fraction of the point defect that leads to the hop.

We obtain
$$\sigma_i = \frac{z_i^2 F^2 \ell^2}{RT} v_0 N x_i (1 - x_i) e^{-\frac{\Delta_m G}{RT}}$$

Writing $\Delta_m G = \Delta_m H - T \Delta_m S$ where $\Delta_m H$ and $\Delta_m S$ denote respectively the variations in enthalpy and entropy that characterize the hop, we obtain

$$\sigma_i = f \frac{z_i^2 F^2 \ell^2}{RT} v_0 N x_i (1 - x_i) e^{-\frac{\Delta_m H - T \Delta_m S}{RT}}$$

or
$$\sigma_i = f \frac{z_i^2 F^2 \ell^2}{RT} v_0 N x_i (1 - x_i) e^{\frac{\Delta_m S}{R}} e^{-\frac{\Delta_m H}{RT}}$$

The results of measurements of ionic conductivity are often represented in the form

$$\sigma_i = \frac{A}{T} e^{-\frac{E_a}{RT}}$$

where A/T is the pre-exponential factor and E_a is the activation energy. By identification, we have

$$A = f \frac{z_i^2 F^2 \ell^2}{R} v_0 N x_i (1 - x_i) e^{\frac{\Delta_m S}{R}} \quad \text{and} \quad E_a = \Delta_m H$$

Note – When $x_i \ll 1$, we often write $N x_i (1 - x) \approx C_i$ where C_i is the defect concentration involved in the ionic displacement.

3.2.3 – Conductivity and temperature

Pure or weakly doped compounds: example of NaCl

The dominant disorder in NaCl is Schottky disorder ($V'_{Na} + V^{\bullet}_{Cl}$). The ionic conductivity as a function of temperature is schematically represented in Arrhenius coordinates in figure 39.

The diagram in figure 39(a) may be separated into four domains whose characteristics are given in table 23.

In domain II, the number of valence of impurities is greater than unity. This domain has the lowest activation energy; namely, that of the migration of Na^+ by a vacancy mechanism. In domain III, the association is due to the formation of complexes (or the precipitation of salts), which involves the enthalpy of association $\Delta_{assoc}H$. In domains I and I', Δ_fH denotes the enthalpy of formation for the Schottky pair.

Table 23 – Characteristics of domains of conduction defined in figure 39.

Regime	Charge carrier	Origin of vacancies	Activation energy
I' intrinsic	Na^+ and Cl^-	thermal	$E_a = f(\Delta_m H_{Na} + \Delta_m H_{Cl} + \Delta_f H)$
I intrinsic	Na^+	thermal	$E_a = \Delta_m H + \dfrac{\Delta_f H}{2}$
II extrinsic	Na^+	impurities	$E_a = \Delta_m H$
III association	Na^+	impurities	$E_a = \Delta_m H + \dfrac{\Delta_{assoc} H}{2}$

In figure 39(b), doping NaCl with $CaCl_2$ is manifested by an increase in conductivity in the extrinsic domain II [from (1) toward (3)] with a shift in the domain toward the high temperatures. The variation in activation energy is not significant, which implies an identical mechanism.

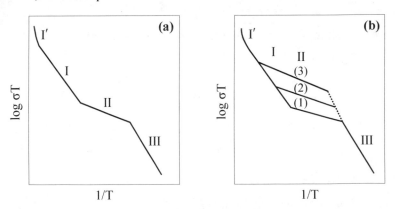

Figure 39 – Arrhenius plot of ionic conductivity in **(a)** pure NaCl and **(b)** NaCl doped by $CaCl_2$, with doping level increasing from (1) to (3).

3.2.4 – Conductivity and environment

We limit ourselves here to the case of an MX_2 crystal, either pure or doped by DX, where the ionic disorder is of the anionic Frenkel type. D substitutes for M in the solid solution MX_2-DX. The expression for particulate conductivity due to a charged species i is given by

$$\sigma_i = z_i \, F \, u_i \, C_i$$

The environment intervenes through the concentration C_i. The expressions for this concentration as a function of partial pressure P_{X_2} are established by following the approach for drawing the Brouwer diagrams. We must, however, account for the significant difference between the ionic mobility u_i and the electronic mobility u_{el} ($u_{el} \approx 10^3 \, u_i$). Figure 40 shows a typical example of the total conductivity σ_t if we concede that the mobility of the charge carriers does not depend on the composition. The horizontal part corresponds to the domain where the ionic crystal behaves as an electrolyte. It defines the electrolytic domain of the crystal. A variation in conductivity with P_{X_2} reflects a mixed conduction (ionic + electronic). The sign of the slope of the curve informs us about the type of semiconductor (n or p). This is to be compared to the theoretical values deduced from the chosen model. In the present case, we arrive at

$$\sigma_n = k\,P_{X_2}^{-\frac{1}{4}} \quad \text{and} \quad \sigma_p = k'\,P_{X_2}^{\frac{1}{4}}$$

for the pure crystal (fig. 41). With a dopant present, we observe a significant increase in the ionic conductivity and an enlargement of the electrolytic domain (fig. 40).

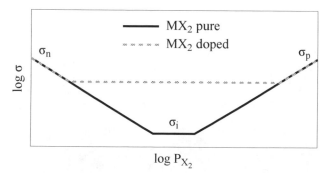

Figure 40 – Schematic plot showing how doping MX_2 by DX influences the total conductivity.

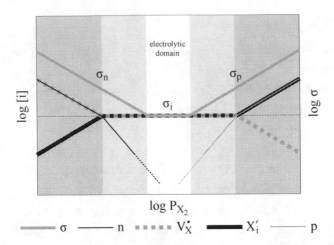

Figure 41 – Concentration of charge carriers (Brouwer diagram) and
total conductivity of pure MX_2 crystal as a function of X_2 partial pressure.

3.2.5 – Ionic conductivity and composition

◇ Case of low doping level (< 1 %)

Research has generally focused on ionic MX-type single crystals. The structure
defects responsible for ionic transport are of extrinsic origin. We generally
observe that ionic conductivity is linear with respect to the doping level, which
gives us access to the electric mobility and to its activation energy. An example
is the solid solution $NaCl$-$MnCl_2$ (fig. 42).

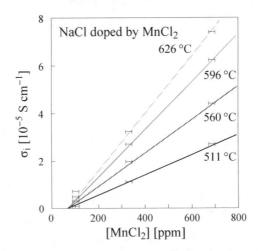

Figure 42 – Ionic conductivity of solid solution $NaCl$-$MnCl_2$
as a function of doping level (from Kirk & Pratt, 1967).

◇ *Case of high doping level (> 1 %)*

This concerns essentially solid oxide solutions that are susceptible of being used as solid electrolytes in electrochemical generators or in gas sensors. We can cite as examples ZrO_2-Y_2O_3, CeO_2-Gd_2O_3, and ThO_2-CaO. These electrolytes are oxide ion conductors. Increasing the doping level leads to a maximum in the ionic conductivity. The presence of this maximum is explained by the simultaneous increase in the concentration of charge carriers and electrostatic interactions between structure defects that inhibit the motion of oxygen. Figure 43 illustrates this behavior for various solid solutions.

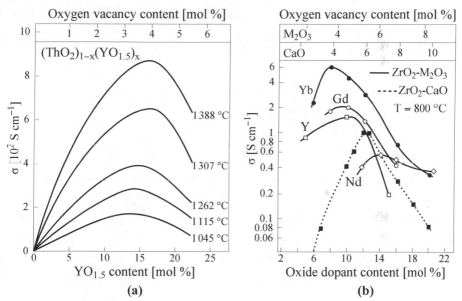

Figure 43 – Ionic conductivity of solid solutions made from (a) ThO_2 (from Hammou, 1975) and (b) ZrO_2 (from Takahashi, 1972).

3.2.6 – Other parameters

Among the other factors that influence the ionic conductivity, we can cite

▷ the crystalline structure, and in particular the layered, cavity, or channel structures (β alumina, MoO_3, $Li_xV_2O_5$, TiS_2),

▷ the size of the oxide cation or of the salt dopant,

▷ the size of the mobile species,

▷ the presence of oxydo-reducible species (transition metals),

▷ aging.

3.3 – Basic description of Wagner theory

The goal is to express the potential difference ΔU between the terminals of a mixed-conduction ionic crystal subjected to a chemical potential gradient from one of its constituents. For this, we consider the following electrochemical chain:

$$X_2(P_a), M/AX/M, X_2(P_b)$$

which involves a mixed conductor (ionic + electronic) AX, two identical metallic electrodes M, and the different partial pressures P_a and P_b of the gas X_2. The following hypotheses are considered:

▷ local equilibrium at each point in the chain,

▷ electroneutrality at each point,

▷ steady state,

▷ negligible overvoltage at the electrodes.

The potential difference is given by

$$\Delta U = \varphi^{(a)} - \varphi^{(b)} = -\frac{1}{F}(\tilde{\mu}_e^{(a)} - \tilde{\mu}_e^{(b)})$$

At each interface, the electrochemical potential of electrons in the metal (e) is the same as that of electrons in the electrolyte (e′). We must now relate these electrochemical potentials to the activity of the various point defects (charged or not) present in the AX compound. For charged structure defects, denoted by $D_i^{z_i \bullet}$, we can write equilibria of the type

$$D_i^{z_i \bullet} + z_i e' \rightleftharpoons D_i^{\times} \quad \text{with } \tilde{\mu}_{D_i^{z_i \bullet}} + z_i \tilde{\mu}_e = \mu_{D_i^{\times}}$$

From the relationships derived above and reproduced here for clarity,

$$\vec{i}_i = z_i F \vec{J}_i \qquad J_i = -C_i \tilde{u}_i \vec{\nabla} \tilde{\mu}_i \qquad \sigma_i = z_i^2 F^2 C_i \tilde{u}_i$$

we can express the total current density i_t

$$\vec{i}_t = \frac{\vec{I}}{S} = \sum_i \vec{i}_i = \sum_i -\frac{\sigma_i}{z_i F} \vec{\nabla} \tilde{\mu}_i$$

where I and S denote the current and the surface area of the electrode-electrolyte contact. Applying this to the equilibrium, which invokes $D_i^{\times}$, gives

$$\vec{i}_t = -\sum_i \frac{\sigma_{ion}}{z_i F} \vec{\nabla} \tilde{\mu}_{D_i^{\times}} + \sum_i z_i \frac{\sigma_{ion}}{z_i F} \vec{\nabla} \tilde{\mu}_e + \frac{\sigma_e}{F} \vec{\nabla} \tilde{\mu}_e - \frac{\sigma_h}{F} \vec{\nabla} \tilde{\mu}_h$$

where σ_{ion} denotes the ionic conductivity.

The electronic equilibrium gives $\vec{\nabla}\tilde{\mu}_e = -\vec{\nabla}\tilde{\mu}_h$,

or
$$\vec{i}_t = -\sum_i \frac{\sigma_{ion}}{z_i F}\vec{\nabla}\mu_{D_i^x} + \frac{1}{F}\left(\sum_i \sigma_{ion} + \sigma_e + \sigma_h\right)\vec{\nabla}\tilde{\mu}_e$$

By writing $\sigma_t = \sum_i \sigma_{ion} + \sigma_e + \sigma_h$,

we obtain
$$\vec{i}_t = -\sum_i \frac{\sigma_{ion}}{z_i F}\vec{\nabla}\mu_{D_i^x} + \frac{\sigma_t}{F}\vec{\nabla}\tilde{\mu}_e$$

which gives
$$\vec{\nabla}\tilde{\mu}_e = \frac{F\vec{i}}{S\sigma_t} + \sum_i \frac{t_{ion}}{z_i F}\vec{\nabla}\mu_{D_i^x}$$

Note that $\vec{\nabla}\tilde{\mu}_e = f(\vec{\nabla}\mu_{D_i^x})$. Because $\vec{\nabla}\mu_{D_i^x}$ is proportional (see example) to $\mu_{X_2} = f(P_{X_2})$, we can use P_{X_2} as the experimental variable to determine the potential difference ΔU between the terminals of the electrochemical chain. Let us establish, for example, the relationship between $\mu_{D_i^x}$ and μ_{X_2} by considering the equilibrium

$$X_X^x \rightleftharpoons V_X^x + \tfrac{1}{2}X_2$$

with here $D_i^x \equiv V_X^x$ and $\mu_{V_X^x} = \mu_{X_X^x} - \tfrac{1}{2}\mu_{X_2}$ and a constant of proportionality $\alpha_i = -\tfrac{1}{2}$.

This gives
$$\vec{\nabla}\tilde{\mu}_e = \frac{F\vec{i}}{S\sigma_t} + \sum_i \frac{t_{ion}\alpha_i}{z_i}\vec{\nabla}\mu_{D_i^x}$$

By considering a gradient along the axis x between b and a, the expression for the potential difference ΔU is $\quad \Delta U = -\dfrac{1}{F}\displaystyle\int_b^a d\tilde{\mu}_e$

$$\Delta U = -\int_b^a \frac{I}{S\sigma_t}dx - \frac{1}{F}\int_b^a \sum_i \frac{t_{ion}\alpha_i}{z_i}d\mu_{X_2}$$

which gives, after integration over a length ℓ

$$\Delta U = -RI - \Delta E$$

where RI represents the ohmic drop across the AX compound

$$R = \frac{1}{\sigma_t} \times \frac{\ell}{S}$$

and ΔE is the potential difference at zero current ($I = 0$).

$$\Delta E = -\frac{1}{F}\int_b^a \sum_i \frac{t_{ion}\alpha_i}{z_i}d\mu_{X_2}$$

Taking the average value of the ionic transport number between the extremities a and b of the chain, we obtain

$$\Delta E = -\frac{\bar{t}_{ion}\alpha_i}{z_i F}\int_b^a d\mu_{X_2}$$

For $\bar{t}_{ion} = 1$ we have a solid electrolyte with

$$\Delta E_{th} = -\frac{\alpha_i}{z_i F} \int_b^a d\mu_{X_2}$$

where ΔE_{th} is the Nernst thermodynamic voltage.

We deduce the relationship $\Delta E = \bar{t}_{ion} \Delta E_{th} = (1 - \bar{t}_e) \Delta E_{th}$

where $\bar{t}_e$ represents the average electronic transport number.

Thus, when AX is a mixed conductor, the potential difference measured between the terminals is less than the thermodynamic voltage. Furthermore, matter is transported from the compartment where the chemical potential of X_2 is highest toward the other compartment. This phenomenon is referred to as electrochemical semipermeability.

Note – This relationship is applied to measure thermodynamic quantities and transport numbers. It also finds use in potentiometric gas sensors, fuel cells, and electrolyzers.

Note – In the presence of an electrochemical semipermeability flux, Wagner relationship cannot be verified experimentally. This may be due to the fact that the activity of species X_2 at the contact surface differs from its activity in the gas phase. A slow desorption kinetics may be at the origin of these problems.

Example of application – We consider a CeO_{2-x} membrane subjected to an O_2 chemical potential gradient that goes from 0.21 bar (compartment a) to 10^{-20} bar (compartment b) with an ionic transport number of 0.7 at 900 K. CeO_{2-x} is a mixed conductor at low oxygen partial pressure.

The reaction at the electrodes is

$$O_O^\times \rightleftharpoons \tfrac{1}{2}O_2 + V_O^{\bullet\bullet} + 2e' \quad \text{or} \quad O_O^\times \rightleftharpoons \tfrac{1}{2}O_2 + V_O^\times$$

Here, $z_i = 2$ and $\alpha_i = -\tfrac{1}{2}$ with $\mu_{O_2} = \mu_{O_2}^0 + RT \ln P_{O_2}$.

The expression for the potential difference is

$$\Delta E = \frac{RT}{4F} \times t_i \times \ln \frac{P_{O_2}^{(a)}}{P_{O_2}^{(b)}}$$

Evaluating this numerically gives

$$\Delta E = \frac{8.314 \times 900}{4 \times 96480} \times 0.7 \times \ln \frac{0.21}{10^{-20}} = 0.635 \, V$$

Note – Other approaches of the Wagner theory are proposed in the literature.

Exercises

Exercise 3.1 – Influence of geometric factor

Table 24 presents the results of complex impedance spectroscopy measurements of several cylindrical samples of solid solution $(CeO_2)_{0.89}(CaO)_{0.11}$ at 263 °C in air. The values correspond to the intersection of the impedance diagram with the real axis in the Nyquist representation and for the high-frequency domain.

Table 24 – Resistance of several cylindrical samples of solid solution $(CeO_2)_{0.89}(CaO)_{0.11}$ at 263 °C in air.

Length ℓ [cm]	Diameter φ [cm]	Resistance R_1 [kΩ]
0.2	0.6	54.20
0.2	1	19.53
0.3	0.8	45.8
0.4	0.5	156.21
0.6	0.8	91.55

1. Draw the curve of the resistance R_1 as a function of the geometric factor k.

2. Derive the equation for $R_1(k)$. What can we conclude?

3. Deduce the electrical conductivity of the solid solution $(CeO_2)_{0.89}(CaO)_{0.11}$ under the given experimental conditions.

4. Impedance measurements were made on several cylindrical samples with differing geometric factors of the solid solution $(CeO_2)_{0.92}(CaO)_{0.08}$ at 521 °C under oxygen. The results show that the resistance recorded at low frequencies is a linear function of the area S of the cylinder cross section and varies with oxygen partial pressure. What conclusion can we make?

Exercise 3.2 – Study of oxygen mobility in solid solutions $(ThO_2)_{1-x}(YO_{1.5})_x$

1. Based on the data in table 25 for the solid solution $(ThO_2)_{1-x}(YO_{1.5})_x$, draw the curve of log u as a function of inverse absolute temperature, where u denotes the ionic mobility.

Table 25 – Electric mobility of oxide ion vacancies in solid solution $(ThO_2)_{1-x}(YO_{1.5})_x$ for several temperatures.

Temperature T [°C]	876	1 166	1 678
Mobility log u [$cm^2 V^{-1} s^{-1}$]	– 4.96	– 4	– 3

Verify that the equation for mobility u has the form $u = u_0 e^{-\frac{E_a}{RT}}$.

2. Calculate E_a in $kJ\,mol^{-1}$.

3. What does E_a represent?

4. **a.** The lattice parameter a of the cubic fluorite-type lattice of a solid solution is determined by X-ray diffraction to be 5.595 Å at ambient temperature. Given that the most likely hopping direction is along the axes of the lattice, determine the most likely hopping distance.

   b. Given that the migration process can be described by the activated hopping model, determine the lattice-vibration frequency v_0 at 876 °C.

Exercise 3.3 – Study of electronic conductivity in solid solutions $(CeO_2)_{1-x}(CaO)_x$

The solid solution $(CeO_2)_{1-x}(CaO)_x$ with $x = 2.8 \times 10^{-3}$ crystallizes in a cubic system with a fluorite-type structure. The dominant disorder is anionic Frenkel disorder.

1. Based on the data in table 26,
   a. draw in logarithmic coordinates the electronic conductivity σ_e as a function of oxygen partial pressure. Limit the range to $10^{-5} \leq P_{O_2} \leq 10^{-2}$ bar and assume that any variation in ionic conductivity with oxygen partial pressure is negligible.
   b. give the equation of $\log \sigma_e = f(\log P_{O_2})$.

Table 26 – Total conductivity σ_t
as a function of oxygen partial pressure P_{O_2}.

P_{O_2} [bar]	1	0.21	10^{-2}	10^{-3}	5.10^{-4}	10^{-4}	10^{-5}
$10^3 \, \sigma_t$ [S cm^{-1}]	1.096	1.096	1.38	1.66	1.79	2.26	3.14

2. What type (n or p) of electronic conduction do we observe in the P_{O_2} range studied?

3. For $P_{O_2} = 10^{-4}$ bar, calculate
   a. the electronic conductivity,
   b. the ionic transport number t_i.

4. Assuming that the electronic mobility u_e is independent of oxygen partial pressure,
   a. give the equation describing the experimental variation $\sigma_e = f(P_{O_2})$,
   b. give a theoretical interpretation of the equation for low P_{O_2}. Give the equilibria and the other necessary relationships required to derive this equation.

Exercise 3.4 – Electronic transport number in a glass

To determine the electronic transport number of a Ag^+-conducting glass, we implement the following electrochemical chain:

$$\text{mixture Hg + Ag / electrolyte / mixture Hg + }\varepsilon\text{Ag}$$
$$\text{compartment 1} \qquad\qquad \text{compartment 2}$$

where the electrolyte is a glass that conducts by Ag^+ ions and where ε denotes a small quantity.

1. Express the electromotive force ΔE at the terminals of this chain as a function of the activities $a_{Ag}^{(1)}$ and $a_{Ag}^{(2)}$ of silver in compartments 1 and 2 and of the average electronic transport number $\bar{t}_e$ of the solid electrolyte. In what follows, we consider that $a_{Ag}^{(2)}$ remains much, much smaller than unity and that $a_{Ag}^{(1)}$ remains constant and equal to unity.

2. We observe that, in the limit of the measurement precision, the electronic transport number is close to zero ($t_e \approx 0$). We now want to determine the transport number more precisely. For this, we follow the open-circuit variation of ΔE as a function of time t.

a. What phenomenon is responsible for this variation? Give a simple explanation for what happens.

b. Give the expression for the derivative $d\Delta E/dt$ as a function of the molar fraction of silver, $x_{Ag}^{(2)}$, in compartment 2 and its derivative $dx_{Ag}^{(2)}/dt$. Assume that the coefficient for the activity of silver in compartment 2 is constant.

3. Derive the expression for the electronic conductivity σ_e of the glass as a function of ΔE, $d\Delta E/dt$, $x_{Ag}^{(2)}$, ℓ, S, and n_T, where ℓ and S represent respectively the thickness and the area of the sample and n_T is the total number of moles of metallic atoms in compartment 2.

4. Compartment 2 contains a 10 mm^3 drop of mercury. The electromotive force is 850 mV. We observe a drift of 10 mV per month. The atomic fraction of silver is less than 6×10^{-4}, which is its saturation value.

a. What is the upper limit of the electrical resistance?

b. What is the upper limit of the electronic transport number of the glass given that the total conductivity is $2.2 \times 10^{-3} \text{ S cm}^{-1}$? Assume $\ell/S = 0.1 \text{ cm}^{-1}$.

Data

> density of mercury: $\rho_{Hg} = 13\,600 \text{ kg m}^{-3}$
>
> molar mass of mercury: $M_{Hg} = 200 \text{ g mol}^{-1}$

Exercise 3.5 – Electrical properties of potassium chloride KCl

1. a. Give the reaction for doping KCl by $BaCl_2$ given that
   ▷ the dominant disorder in KCl is Schottky pair,
   ▷ barium substitutes for potassium.

   b. Show that the doping leads to an increase in the cationic transport number t_+. Neglect the mobility of Ba^{2+} ions.

   c. The solid solution obtained after doping has a NaCl-type structure. Denoting its formula by $(KCl)_{1-x}(BaCl_2)_x$, calculate the concentration in mol cm^{-3} and in mol L^{-1} of the potassium vacancies $[V_K']$ for $x = 1.4 \times 10^{-4}$. Given the low value of x, we neglect the influence of $BaCl_2$ on the density and on the molar mass.

Data

> density of KCl: 1.984 g cm^{-3}
>
> molar mass of KCl: 74.56 g mol^{-1}

2. Starting from the curve shown in figure 44 and using the Nernst-Einstein relation, calculate the electrochemical mobility and the diffusion coefficient of V'_K at 440 °C and give their units. Assume that all potassium vacancies of extrinsic origin participate in the conduction.

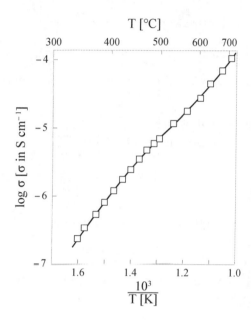

Figure 44 – Electrical conductivity of solid solution $(KCl)_{1-x}(BaCl_2)_x$ with $x = 1.4 \times 10^{-4}$ in Arrhenius coordinates.

3. From the literature, the diffusion coefficient of K^+ in KCl, which we denote by $D*_K$, is given by $D^*_K = 2 \times 10^{-5} e^{-\frac{0.76\,eV}{kT}}$ where $D*_K$ is in $cm^2 s^{-1}$.
   a. Calculate the diffusion coefficient at 440 °C and compare it with that of V'_K found above in question 2.
   b. Show that the relationship $D^*_K [K^\times_K] = D_{V_K}[V'_K]$ is approximately verified.
   c. Show that the activation energy of the diffusion D^*_K of potassium vacancies equals that of the conductivity.
   d. Deduce the enthalpy of migration of the potassium vacancies V'_K given in $kJ\,mol^{-1}$.

4. Assume in what follows that the mobility of potassium vacancies V'_K does not depend on the doping level x (which remains low). Given that the self-diffusion coefficient of potassium in a pure KCl crystal takes the form

$$D^*_K = 4 \times 10^{-2} e^{-\frac{1.48\,eV}{kT}} \qquad \text{where } D^*_K \text{ is in } cm^2 s^{-1},$$

   a. calculate the concentration of potassium vacancies, V'_K, in a pure crystal at 440 °C;

   b. deduce the cationic conductivity;

   c. calculate the enthalpy of formation of Schottky pair.

5. The self-diffusion coefficient for chlorine in pure KCl is given by

$$D^*_K = 200\,e^{-\frac{2.24\,eV}{kT}} \quad \text{where } D^*_K \text{ is in } cm^2\,s^{-1}.$$

   a. Calculate the mobility and enthalpy of migration of chlorine vacancies $V^{\bullet}_{Cl}$ at 440 °C.

   b. Determine the anionic transport number t_{Cl^-} at 440 °C.

   c. At what temperature does the transport number approach 0.5?

Data
| Melting temperature of KCl: 776 °C

6. *A priori*, which point defect would induce LiCl doping of solid KCl and how would this affect the conductivity?

Exercise 3.6 – Application of Nernst-Einstein relation to LiCF$_3$SO$_3$ in poly(ethylene oxide) P(EO)

The ionic conductivity σ of the polymer P(EO)$_{36}$-LiCF$_3$SO$_3$, measured by complex impedance spectroscopy at 358 K is $9 \times 10^{-5}\,S\,cm^{-1}$. The diffusion coefficient of lithium, measured at the same temperature, is $1.53 \times 10^{-6}\,cm^2\,s^{-1}$. Assume that the salt LiCF$_3$SO$_3$ disassociates according to the following balanced reaction:

$$LiCF_3SO_3 \rightleftharpoons Li^+ + CF_3SO_3^-$$

Note – P(EO): poly(ethylene oxide), LiCF$_3$SO$_3$: lithium trifluoromethanesulfonate

1. Calculate the number of lithium ions that partake in electric conduction given that the transport number of lithium is 0.45.

2. By using the approximation given below, deduce the disassociation rate for the salt.

Data
| Mass of sample of P(EO)$_{36}$-LiCF$_3$SO$_3$: 48.33 g

Element	H	C	O
Molar mass $[g\,mol^{-1}]$	1	12	16

$M(LiCF_3SO_3) = 156$ g mol^{-1}

Density of $P(EO)_{36}$-$LiCF_3SO_3 = 1.3$ g cm^{-3}

Exercise 3.7 – Electrical conductivity as a function of composition in $(CeO_2)_{1-x}(YO_{1.5})_x$

Table 27 presents the experimental results for electronic conductivity σ_e of the solid solution $(CeO_2)_{1-x}(YO_{1.5})_x$ measured at 793 °C under an oxygen partial pressure of 1.5×10^{-13} bar and as a function of doping level x.

Table 27 – Electronic conductivity of solid solution $(CeO_2)_{1-x}(YO_{1.5})_x$ as a function of doping level.

Doping level x	1.5×10^{-1}	10^{-1}	10^{-2}	10^{-3}	5×10^{-4}
σ_e [S cm^{-1}]	3.95×10^{-3}	4.85×10^{-3}	1.26×10^{-2}	3.42×10^{-2}	4.63×10^{-2}

1. Draw the curve for the function $\sigma_e = f(x)$ in logarithmic coordinates.

2. Derive the relation $\log \sigma_e = a \log x + b$ where a and b are constants to be determined.

3. The dominant disorder in CeO_2 is Frenkel anionic disorder. In the solid solution $(CeO_2)_{1-x}(YO_{1.5})_x$, the yttrium substitutes for cerium. Given that

   ▷ the dominant electronic conductivity is n type,

   ▷ the electronic mobility is constant, and

   ▷ the species responsible for the ionic conduction is the oxide ion by a vacancy mechanism,

   establish the theoretical equation for the electronic conductivity as a function of doping level x.

Exercise 3.8 – Conductivity of nickel oxide

At high oxygen partial pressure, nickel oxide NiO is a p-type semiconductor. The total electrical conductivity of NiO for various oxygen partial pressures is given in table 28 for a temperature of 1 000 °C.

Table 28 – Conductivity of NiO as a function of oxygen partial pressure at 1 000 °C (from Pope & Birks, 1977).

P_{O_2} [bar]	$10^2 \sigma$ [S cm^{-1}]	P_{O_2} [bar]	$10^2 \sigma$ [S cm^{-1}]	P_{O_2} [bar]	$10^2 \sigma$ [S cm^{-1}]
1	16.7	4.79×10^{-4}	2.66	4.79×10^{-9}	0.266
0.21	10.0	1.58×10^{-4}	1.87	4×10^{-9}	0.251
1×10^{-2}	5.55	1.32×10^{-4}	1.80	1.91×10^{-9}	0.213
1.92×10^{-2}	5.95	1.23×10^{-4}	2.41	1.59×10^{-9}	0.227
1.33×10^{-2}	5.41	1.10×10^{-7}	0.593	1.14×10^{-9}	0.208
2.52×10^{-2}	3.77	3.30×10^{-7}	0.427	1.1×10^{-9}	0.200
1.20×10^{-3}	3.39	1.00×10^{-8}	0.300	4.49×10^{-10}	0.192
6.93×10^{-4}	2.22	5.17×10^{-9}	0.282	2.47×10^{-10}	0.175

1. In logarithmic coordinates, graph the total electrical conductivity of NiO as a function of oxygen partial pressure at 1 000 °C. Determine the slopes of the lines at low ($P_{O_2} < 10^{-6 \text{ bar}}$) and high ($P_{O_2} > 10^{-4 \text{ bar}}$) pressure.

2. Describe the theoretical models for defects that explain the slopes observed.

3. Explain the different behavior as a function of oxygen partial pressure. What are the dominant atomic defects at the highest P_{O_2}?

4. Another study on the conductance G of NiO gives the results shown in figure 45. Given the results of the preceding questions, comment on the slopes of the lines.

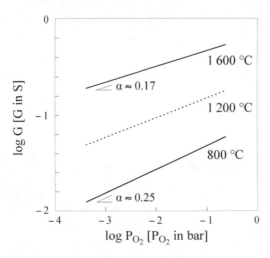

Figure 45 – Logarithm of conductance of a sample of NiO as a function of oxygen partial pressure for various temperatures (from Pope & Birks, 1977).

Exercise 3.9 – Ionic conductivity-activity relationship of glass modifier in oxide-based glasses

1. In $(SiO_2)_{1-\alpha}(Na_2O)_\alpha$-type glasses, identify the glass former and the glass modifier. Give two examples of glasses that involve other formers and modifiers.

2. What is the difference between glass former and modifier in terms of the metal-oxygen bond?

3. Give the coordination indices for silicon and oxygen in these solids.

4. Consider the following electrochemical chain:

$$(Hg,Na)_1 / (SiO_2)_{1-\alpha}(Na_2O)_\alpha / (Hg,Na)_2$$

   (Hg,Na) denotes an amalgam of mercury and sodium. Given that the electrical conductivity in $(SiO_2)_{1-\alpha}(Na_2O)_\alpha$ is purely cationic,
   a. write the expression for the potential difference ΔE measured between the terminals of the chain as a function of molar fraction x_{Na} of sodium in the amalgam,
   b. determine its value at 200 °C for $x_{Na}^{(1)} = 2 \times 10^{-3}$ and $x_{Na}^{(2)} = 10^{-5}$.

5. Consider a cell where the initial amount of amalgam is 10^{-2} mol in compartments 1 and 2. What current I in mA must pass through the chain in 1 min for the potential difference between the idle terminals to be zero. Neglect any variation in the amount of amalgam.

6. The impedance diagram (Nyquist) relative to the volume properties of the glass $(SiO_2)_{1-\alpha}(Na_2O)_\alpha$ is a semi-circle passing through the origin. How does this diagram change once the glass is completely crystallized by thermal treatment?

Exercise 3.10 – Electrochemical coloration

A single crystal of potassium bromide KBr is placed in an atmosphere where the bromine (Br_2) partial pressure is 10^{-10} bar, independent of temperature. The dominant intrinsic disorder in KBr is Schottky disorder. The equilibrium constant K_S for defect formation and the ionic diffusion coefficients are given by the following expressions:

$$K_S = K°e^{-\frac{2.2\,eV}{kT}} \tag{1}$$

$$D_K = D°_K e^{-\frac{1.26\,eV}{kT}} \quad \text{where } D°_K = 10^{-2}\,cm^2\,s^{-1} \tag{2}$$

$$D_{Br} = D°_{Br} e^{-\frac{2.61\,eV}{kT}} \quad \text{where } D°_{Br} = 3 \times 10^4\,cm^2\,s^{-1} \tag{3}$$

1. Give the literal expressions for the partial ionic conductivities as a function of the parameters in the expressions (1) to (3). Assume that the concentrations n and p of the electronic species e′ and h•, respectively, are negligible.

2. Calculate the ratio of partial ionic conductivities, σ_K / σ_{Br}, at 627 °C and deduce the transport number t_K of potassium given that the electronic transport number is negligible.

3. Based on the Brouwer diagram for KBr at 627 °C in figure 46 and assuming that the electrochemical mobility of holes (h•) is very close to that of potassium vacancies at the given temperature, determine the order of magnitude of the electronic transport number t_h.

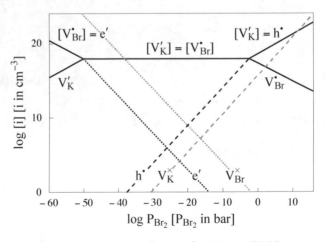

Figure 46 – Brouwer diagram for KBr at 627 °C.

4. In another experiment, we consider a KBr single-crystal tablet that makes an airtight separation between two atmospheres whose bromine partial pressures P_{Br_2} differ. To make the electrodes, each face of the tablet is covered by a layer of inert metal. At the electrode that is used as the anode in what follows, the bromine partial pressure P_{Br_2} is held constant at 10^{-10} bar. With

the anode serving as reference electrode, the potential difference measured between the two electrodes is -1 V in these conditions.

a. What is the partial pressure in the cathodic compartment before any current flows?

b. After the current flows, the single crystal undergoes a blue coloration near the cathode, whereas it was perfectly transparent prior to current flow.

▷ Write the reactions of the anodic and cathodic electrodes.

▷ What is the origin of the coloration?

c. When the current is stopped, the residual potential difference of -2.5 V is observed. What is the proportion of color centers in the single crystal with respect to the anionic vacancies?

5. In a final experiment, we introduce bromine at $P_{Br_2} = 1$ bar into one compartment and in the other we introduce potassium vapor at a partial pressure of $P_K = 10^{-4}$ bar.

a. Does the crystal turn blue? If it does, near what compartment does it turn blue? Give the concentration of the majority species at each interface when the interfaces are at equilibrium.

Data

> Standard free enthalpy of dissociation of solid KBr in gaseous bromine Br_2 and in gaseous potassium at 627 °C: $\Delta_d G^\circ_{627\,°C} = 327.4$ kJ mol^{-1}

b. What macroscopic phenomenon should we observe over a large timescale if the two compartments are closed and have a small volume?

c. Give a qualitative summary of the displacement of the various charges.

d. How does the measured potential difference evolve?

Exercise 3.11 – Oxygen diffusion in gadolinia-doped ceria

The diffusion coefficient D of oxygen in gadolinia-doped ceria ($Ce_{0.7}Gd_{0.3}O_{2-\delta}$) is given in table 29 for various temperatures.

Table 29 – Diffusion coefficient for oxygen
in solid solution $Ce_{0.7}Gd_{0.3}O_{2-\delta}$ at various temperatures.

T [°C]	569	639	712	813	863	951
D [cm^2 s^{-1}]	3.5×10^{-9}	1.8×10^{-8}	6×10^{-8}	1.2×10^{-7}	2.1×10^{-7}	4.5×10^{-7}

1. Write the equation for cerium substituted by gadolinia in ceria. What is the point of this substitution?

2. **a.** Graph the product DT as a function of temperature in Arrhenius coordinates.
   b. Determine the activation energy for ionic diffusion in gadolinia-doped ceria in $kJ\,mol^{-1}$ and in eV.

3. What are the units of electric mobility of a charge carrier?

4. Assuming that 50% of the vacancies are mobile, calculate the diffusion coefficient, the electric mobility, and the conductivity at 800 °C.

Data

   Lattice parameter for ceria in the temperature range considered
   $a = 5.426$ Å.

Exercise 3.12 – Electrical conductivity of solid vitreous solution $(SiO_2)_{1-x}(Na_2O)_x$

Consider the following electrochemical chain:

$$O_2(P),Pt/(1-x_1)\,SiO_{2-x_1}M_2O/(1-x_2)\,SiO_{2-x_2}M_2O/Pt,O_2(P)$$
$$\langle\alpha\rangle \qquad\qquad \langle1\rangle \qquad\qquad\qquad \langle2\rangle \qquad\qquad \langle\alpha'\rangle$$

in which the phases $\langle1\rangle$ and $\langle2\rangle$ are fused silica glass SiO_2 with different fractions x_1 and x_2 of an alkali metal oxide M_2O. P denotes the oxygen partial pressure, which is the same at both terminals of the chain. The phases $\langle\alpha\rangle$ and $\langle\alpha'\rangle$ consist of platinum Pt in contact with oxygen.

1. Given that the glasses are conducting due to the M^+ cations, express the voltage between $\langle1\rangle$ and $\langle2\rangle$ as a function of the chemical potential of the M^+ ions in $\langle1\rangle$ and $\langle2\rangle$.

2. Assuming that the temperature is sufficiently high to establish the equilibrium
   $$\tfrac{1}{2}\,O_{2(g)} + 2e_{\langle Pt\rangle} \rightleftharpoons O^{2-}_{(glass)}$$
   which we may also write in the form

   $$\tfrac{1}{2}O_{2(g)} + 2M^+ + 2e_{\langle Pt\rangle} \rightleftharpoons M_2O$$

   give the expression for the emf ΔE at the chain terminals as a function of the thermodynamic activity a_{M_2O} of the alkali-metal oxide in $\langle1\rangle$ and $\langle2\rangle$.

3. Based on the "weak electrolyte" model applied to amorphous ionic conductors, relate the emf to the conductivity of the cations M^+ in $\langle 1 \rangle$ and $\langle 2 \rangle$. Assume that the mobilities are identical and independent of concentration.

4. The measurements made at various temperatures reveal a linear relationship between ΔE and T. The extrapolation of the curve $\Delta E(T)$ intercepts the ordinate axis at 100 mV. Given that the equation for the conductivity as a function of temperature is

$$\sigma = \frac{\sigma_0}{T} e^{-\frac{E_a}{RT}}$$

where σ_0 is a constant and E_a is the activation energy, what is the difference ΔE_a between activation energies of conduction (expressed in eV and in $kJ\,mol^{-1}$) in the two vitreous phases?

Exercise 3.13 – High-temperature protonic conductor SrZrO$_3$

A – Point defects in the presence of oxygen

1. In the absence of water vapor, stoichiometric $SrZrO_3$ exhibits Schottky disorder. Write the internal equilibrium reactions (ionic and electronic defects) and the expressions for the corresponding equilibrium constants.

2. Write the equilibrium reaction when gaseous oxygen is present in the environment and the expression for the corresponding equilibrium constant.

B – Point defects in the presence of water vapor

In the presence of water vapor, we observe an increase in the electrical conductivity due to the onset of conduction by interstitial protons.

1. Write the equilibrium reaction for $SrZrO_3$ with water vapor and give the expression for the corresponding equilibrium constant.

2. Figure 47 proposes a variation in point-defect concentration with oxygen partial pressure.
   Assuming that, near the sample, the sum of the hydrogen and water partial pressures is constant at 1 bar, show that the concentration of interstitial protons $H_i^{\bullet}$ may be expressed as

$$[H_i^{\bullet}] = \frac{K[V_O^{\bullet\bullet}]^{1/2}}{\left(1 + \dfrac{P_{H_2}}{P_{H_2O}}\right)^{1/2}} \qquad \text{where K is a constant.}$$

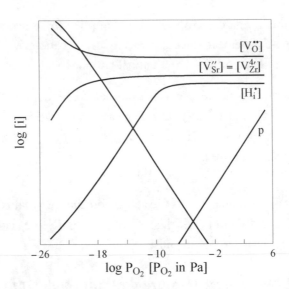

Figure 47 – Concentration of point defects as a function of oxygen partial pressure in the presence of water vapor at 800 °C (from Labrincha, Frade & Marques, 1993).

3. Justify the equation for the variation of $[H_i^\bullet]$ with oxygen partial pressure (see fig. 47).

Data

$$2H_2 + O_2 \rightleftarrows 2H_2O$$

with $\quad \Delta_r G_T^\circ = -494 + 0.112\, T \ [\text{kJ mol}^{-1}]$

4. We dope $SrZrO_3$ with yttrium oxide Y_2O_3 to obtain a solid solution $SrZr_{1-x}Y_xO_{3-0.5x}$.

   a. Write the doping reaction and the equation for electroneutrality.

   b. State the conditions that would allow an oxygen partial pressure domain where $[Y_{Zr}']$, $[V_{Sr}'']$, $[V_{Zr}^{4'}]$, and $[V_O^{\bullet\bullet}]$ are constant.

   c. For these conditions, give the equation that approximately relates $[H_i^\bullet]$ to $[V_O^{\bullet\bullet}]$ and comment on the type of ionic conduction.

C – Determination of electronic conduction part

1. We make the following cell:

$$H_2\, (P_{H_2,1} = 1 \text{ bar}), Pt\, / \, SrZr_{0.95}Y_{0.05}O_{3-\alpha}\, / \, Pt, H_2\, (P_{H_2,c})$$

Tables 30 and 31 give the variation ΔE in the emf for this cell for various hydrogen partial pressures and temperatures, respectively.

$P_{H_2,c}$ [bar]	6×10^{-2}	2×10^{-2}	10^{-2}
ΔE_{expt} [mV]	131	183	214

Table 30 – emf of cell for various hydrogen partial pressures at 800 °C.

T [°C]	600	800	1 000
ΔE_{expt} [mV]	177	214	254

Table 31 – emf of cell at various temperatures for a hydrogen partial pressure of 10^{-2} bar.

a. Derive the equation for the emf as a function of hydrogen partial pressure $(P_{H_2,1} = P_{H_2,c})$ and temperature T.
b. Indicate the conditions for which the equation is valid.
c. Compare the experimental results with theory and draw your conclusions.

2. Another way to estimate the ionic transport number in this conductor is to make the following cell:

$$(+) \ H_2, Pt / SrZr_{1-x} Y_x O_{3-\alpha} / Pt, Ar \ (-)$$

a. Explain how, by applying a continuous current across the cell, we can determine the type of the conduction and the ionic transport number. For two compositions, figure 48 gives the hydrogen flux density traversing the electrolyte, expressed in $mL(NTP) mn^{-1} cm^{-2}$, as a function of current density at 800 °C.

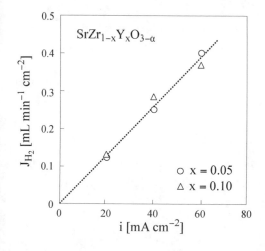

Figure 48 – Hydrogen flux density traversing electrolyte $SrZr_{1-x} Y_x O_{3-\alpha}$ as a function of continuous current density traversing the cell at 800 °C.

b. Calculate the ionic transport number and draw a conclusion.

3. Propose a more precise method involving electrochemical devices to evaluate the electronic part of the conduction for this protonic conductor.

Exercise 3.14 – Free-volume model

1. State the four hypotheses used in developing the free-volume model applied to the variation in ionic conductivity of molten electrolytes and vitrifiable mixtures above their glass transition temperature T_G.

2. Consider an electrolyte whose cationic conductivity is due to the species M^+ (Li^+, Na^+, K^+, Ag^+, ...).
   a. Express the hopping probability P of this species as a function of average free volume $\overline{V}_f$.
   b. Assume that the average free volume $\overline{V}_f$ is zero at the temperature T_0 and, to a first approximation, obeys the relation

   $$\overline{V}_f = V_0 \Delta\alpha (T - T_0)$$

   where $\Delta\alpha$ is the difference between the thermal expansion coefficients of the liquid and the solid, V_0 is the system volume at temperature T_0, and T_0 is the ideal glass transition temperature.
   Express this probability as a function of temperature.
   c. The relationship obtained may be written as

   $$P = e^{-\frac{B}{R(T-T_0)}}$$

   where B is a constant and R is the ideal gas constant.
   Express B as a function of the parameters used in (b) and give its units.

3. By applying the free-volume model, the conductivity σ_+ of a cation obtained from the hopping probability is

$$\sigma_+ = \frac{A}{T} e^{-\frac{V_f^*}{V_0 \Delta\alpha (T-T_0)}}$$

where A is a constant and V_f^* is the minimum free volume required to allow the displacement of cation M^+. Table 32 gives the experimental results for the cationic conductivity of the vitreous system $2SiO_2$-K_2O.

Table 32 – Cationic conductivity of vitreous
system $2SiO_2$-K_2O as a function of temperature.

σ_+ [S cm^{-1}]	T [°C]	σ_+ [S cm^{-1}]	T [°C]	σ_+ [S cm^{-1}]	T [°C]
7.61×10^{-7}	124	5.5×10^{-4}	352	5.66×10^{-2}	652
1.47×10^{-5}	200	5.23×10^{-3}	501	0.138	756
3.20×10^{-5}	227	1.34×10^{-2}	546	0.252	858
1.50×10^{-4}	285	2.88×10^{-2}	600	0.658	1 060

a. ▷ Graph the function $\log \sigma_+ T = f(\frac{1}{T})$ and determine the temperature range over which the function is linear and follows an Arrhenius equation of the form

$$\sigma_+ T = A_1 \, e^{-\frac{E_a}{RT}}$$

- Determine the values of A_1 and E_a.
- Derive the equation for the product $\sigma_+ T$ within the linear regime.

▷ What is the significance of the magnitude of E_a? Write the expression relating E_a to the enthalpies of formation, $\Delta_f H$, and of migration, $\Delta_m H$, of the cationic carrier.

b. ▷ Given the following empirical relationships:

$$T_0 \approx \frac{1}{2} T_f \quad \text{and} \quad T_G \approx \frac{2}{3} T_f$$

where T_f is the melting temperature, show that the non-linear regime of the function $\log \sigma_+ T = f(\frac{1}{T})$ takes a VTF form (Vogel-Tammann-Fulcher)

$$\sigma_+ T = A_2 \, e^{-\frac{B_2}{R(T-T_0)}}$$

- Determine the values of A_2 and B_2.
- Derive the equation for the product $\sigma_+ T$ in the non-linear range.
- Place on a figure the experimental points and the curves corresponding to the theoretical equations for the entire temperature range.

▷ For the non-linear regime, graph the function $\log \sigma_+ T = f\left(\frac{1}{T-T_0}\right)$ with T_0 taken from the previous question. Conclude and find again the values for A_2 and B_2.

▷ Given that the free-volume model applies to the system under consideration, determine the ratio V_f^*/V_0.

Data

$|$ Thermal expansion coefficient of glass: $\alpha_v = 5.58 \times 10^{-5}\,K^{-1}$

Thermal expansion coefficient of liquid: $\alpha_l = 1.455 \times 10^{-4}\ \text{K}^{-1}$

Glass transition temperature of the mix $2SiO_2\text{-}K_2O$: $T_G = 757$ K

4. Given the experimental data, determine the temperature at which the curves corresponding to the two domains cross. Compare this value with T_0 and make your conclusion.

Exercise 3.15 – Study of single-crystal calcium fluoride CaF_2 in the presence of oxygen

Pure calcium fluoride CaF_2 is dominated by anionic Frenkel disorder. The mobility of the calcium ions is assumed to be zero. The electric transport is due to the fluorine vacancies. In the presence of oxygen, oxide ions can enter the CaF_2 structure by substituting for fluorine in the anionic sublattice, which releases gaseous fluorine. No electronic conduction occurs.

1. Give the relevant equilibria.

2. How should we interpret the increase in ionic conductivity?

3. We want to study the conductivity of CaF_2 as a function of temperature and of oxygen partial pressure P_{O_2}. For this, we first measure the impedance under an argon atmosphere ($P_{O_2} = 3 \times 10^{-6}$ bar) of the following cell:

$$O_2, Pt/CaF_2 \text{ (single crystal)}/Pt, O_2$$

The sample consists of a cylinder of length $\ell = 0.816$ cm and diameter $d = 0.69$ cm. Figure 49 shows the impedance diagram obtained at 661 °C.

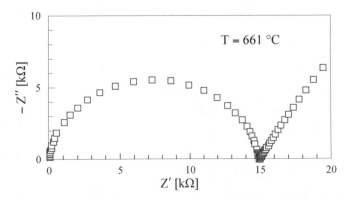

Figure 49 – Impedance diagram for a CaF_2 single crystal obtained under an argon atmosphere at 661 °C (see course notes, chapter 2).

a. Calculate the conductivity of CaF_2 at 661 °C under an argon atmosphere.

b. The evolution of the conductivity under an argon atmosphere and as a function of temperature is shown in figure 50.

Three temperature ranges are apparent: A: T > 653 °C, B: 477 °C < T < 653 °C, and C: T < 477 °C.

▷ Calculate the activation energy (expressed in eV) in ranges A and B.

▷ Deduce the migration energy of fluorine vacancies in CaF_2 and the formation energy of a pair of intrinsic defects. Justify your answer by giving the expression for the conductivity as a function of temperature in each temperature range.

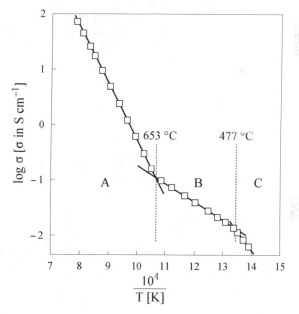

Figure 50 – Conductivity of CaF_2 single crystal under argon atmosphere plotted in Arrhenius coordinates (from Levitskii *et al.*, 1976).

c. The study of the conductivity of CaF_2 as a function of oxygen partial pressure in the temperature range B gives an empirical $P_{O_2}^{1/4}$ relationship. Propose a model that takes into account this result and explain the hypotheses used. Take the mobility of the fluorine vacancies to be independent of their concentration.

Exercise 3.16 – emf of a membrane crossed by an electrochemical semipermeability flux

We want to establish the emf of an electrochemical cell consisting of a membrane traversed by an electrochemical semipermeability flux. The study involves the following setup (fig. 51):

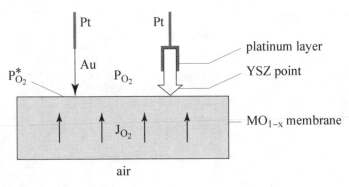

Figure 51 – Schematic of experimental setup.

The electrochemical chain, with the interfaces indicated, is given below:

$$Pt,I \: / \: Au \: / \: MO_{1-x} \: / \: YSZ \: Point \: / \: Pt,O_2 \: / \: Pt,II$$
$$\alpha \quad\;\; \beta \quad\quad \chi \quad\quad\quad\quad \delta \quad\quad\; \varepsilon$$

The membrane consists of a majority electronic conductor non-stoichiometric compound MO_{1-x}, denoted MO in the following. It carries an electrochemical semipermeability flux that results in a departure from equilibrium at the surface where the gold points and the stabilized zirconia are placed. In these conditions, we can consider that the oxygen activity at the oxide surface results in a steady state linked to the reaction

$$O^{2-}_{MO} \longrightarrow \tfrac{1}{2}O_{2\,(surface)} + 2e_{MO}$$

The oxygen partial pressure at the surface, denoted $P^*_{O_2}$, differs from that of oxygen in the gas. This pressure $P^*_{O_2}$ is considered to be in equilibrium with the membrane.

1. Show that the relationship

$$\tfrac{1}{2}\mu^*_{O_2} + 2\mu_e^{MO} = \mu^{MO}_{O^{2-}}$$

holds under these conditions.

2. In the setup used, the gold serves as an inert probe; that is, a purely electric (as opposed to electrochemical) probe. Write the equilibrium involving the electrons present in the gold and in the non-stoichiometric compound.

3. The use of a stabilized zirconia micro-point allows us to consider in a first approximation that no flux crosses the interface MO_{1-x}/YSZ. Given these conditions, write the expression relating the electrochemical potentials of the species O^{2-} in MO_{1-x} and in YSZ.

4. Assuming that the MO_{1-x} surface is an equipotential, draw a qualitative graph of the potential across the chain.

5. What type of conductivity can we determine based on measurements of a semipermeability flux?

6. Establish the expression for the emf ΔE of the cell as a function of $P_{O_2}^*$ and P_{O_2}.

Exercise 3.17 – Determination of electronic conductivity by electrochemical semipermeability

The setup shown below (fig. 52) is implemented to evaluate the electronic conductivity across a majority ionic conducting membrane by measuring the electrochemical semipermeability flux.

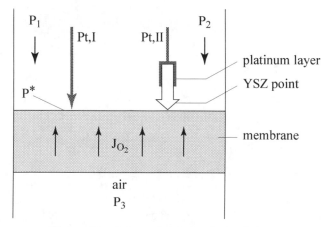

Figure 52 – Schematic of experimental setup.

The membrane consists of stabilized zirconia with 9% yttrium oxide. The electrochemical chain is

$$\text{Pt,I / Oxide electrolyte / YSZ Point / Pt,O}_2 \text{ / Pt,II}$$
$$\alpha \qquad\qquad\qquad \beta \qquad\qquad \gamma \qquad \delta$$

The oxygen partial pressure is denoted P_1 at the entrance, P_2 at the exit, P^* at the membrane surface, and P_3 on the air side (fig. 52).

1. According to the hypotheses of the Wagner theory, the membrane surfaces are in equilibrium with the gaseous environment. Recall the expression of the semipermeability flux as a function of partial pressures P_2 and P_3 on one side and the other of the membrane.

2. Figure 53 represents the semipermeability flux as a function of $P_3^{1/4} - P_2^{1/4}$. We find that the relationship is not linear, which indicates that the hypothesis of the Wagner theory is not valid.

   By accounting for the flux of matter, show that the membrane surface is not in equilibrium with the gas; in particular on the side of low oxygen partial pressure.

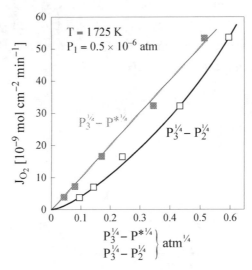

Figure 53 – Semipermeability flux measured as a function
of $P_3^{1/4} - P^{*1/4}$ and of $P_3^{1/4} - P_2^{1/4}$ (from Fouletier, Fabry & Kleitz, 1976).

3. Oxygen activity $a_{O_2}^*$ in the upper layer of the membrane is identified with an oxygen pressure $P_{O_2}^*$. The shape of the stabilized zirconia micro-point

(YSZ) implies that the oxygen flux does not reach the electrode consisting of the platinum layer on the YSZ micro-point. We thus consider that the platinum electrode is in equilibrium with the environmental pressure P_{O_2}. The platinum micro-electrode in contact with the oxide measures the oxygen activity at the oxide surface. Qualitatively graph the potential across the electrochemical chain.

4. The partial pressure of the gas in contact with the points is fixed at 0.5×10^{-6} bar. Determine the pressure $P_{O_2}^*$ given that the potential difference measured between the two platinum wires is 40 mV at 1 452 °C.

5. Taking into account the oxygen activity at the surface of the membrane, the semipermeability flux at the surface of membrane P* is linear (fig. 53). Based on this figure, evaluate the oxygen flux at 1 425 °C if $P_3 = 0.2$ bar and $P_1 = 0.5 \times 10^{-6}$ bar.

6. Give the expression for semipermeability flux as a function of electronic conductivity of the electrolyte on each side of the membrane if P_3 is much, much greater than P_2.

7. Evaluate the electronic conductivity for a 4-mm-thick tablet.

Solutions to exercises

Solution 3.1 – Influence of geometric factor

1. The curve for the resistance R_1 as a function of the geometric factor is shown in figure 54 from the data in table 33.

Table 33 – Resistance of several cylindrical samples of solid solution $(CeO_2)_{0.89}(CaO)_{0.11}$ for various geometric factors.

Geometric factor k [cm^{-1}]	0.255	0.597	0.707	1.194	2.037
Resistance R_1 [kΩ]	19.53	45.80	54.20	91.55	156.21

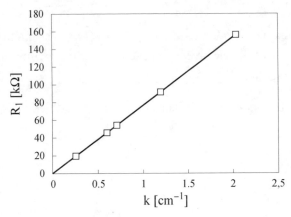

Figure 54 – Resistance R_1 as a function of geometric factor.

We find a linear variation in resistance.

2. The equation for the line in figure 54 is

$$R = 76.594 \times k$$

we conclude that R_1, which was obtained at high frequency, is representative of the ohmic resistance of the solid solution. It is a bulk property.

3. The conductivity σ of the solid solution is given by

$$\sigma = \frac{1}{R} \times \frac{\ell}{S} = \frac{k}{R}$$

Numerical evaluation gives

$$\sigma = \frac{k}{R} = \frac{1}{76.594 \times 10^3}$$

$$\sigma = 1.3 \times 10^{-5}\,\text{S}\,\text{cm}^{-1}$$

4. The circular arc recorded at low frequencies is representative of a surface phenomenon. In this case it is the reaction at the electrode involving the oxygen partial pressure and can be written in the form

$$\tfrac{1}{2}O_{2(g)} + 2e' + V_{\ddot{O}} \rightleftharpoons O_{O}^{\times}$$

Solution 3.2 – Study of oxygen mobility in solid solutions $(ThO_2)_{1-x}(YO_{1.5})_x$

1. From the data in table 34, we can plot the curve for log u as a function of inverse absolute temperature (fig. 55).

Table 34 – Logarithm of mobility u of oxide ion vacancies for several values of inverse temperature.

T [°C]	876	1166	1678
T [K]	1149	1439	1951
$\dfrac{10^4}{T[K]}$	8.7	6.95	5.13
log u [cm^2 V^{-1} s^{-1}]	– 4.96	– 4	– 3

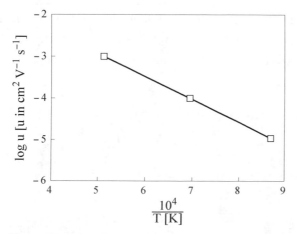

Figure 55 – Electric mobility in Arrhenius coordinates.

We find that this function is linear. It can be expressed in the form

$$u = u_0 e^{-\frac{E_a}{RT}}$$

2. To calculate E_a, we use the two pairs (u_i, T_i) from table 35.

Table 35 – Log of mobility for two temperatures.

Temperature T [°C]	1 166	1 678
log u [cm^{2}V^{-1}s^{-1}]	– 4	– 3

Given that $\log u = f(\frac{1}{T})$ is linear, we can easily find

$$E_a = -R\frac{T_1 T_2}{T_2 - T_1} \ln\frac{u_1}{u_2}$$

Numerical evaluation gives

$$E_a = 105 \text{ kJ mol}^{-1}$$

By using the mobility measured at 1 678 °C, we obtain

$$u_0 = 0.647 \text{ cm}^2 \text{V}^{-1} \text{s}^{-1}$$

The mobility as a function of temperature is

$$u = 0.647 \, e^{-\frac{105\,[\text{kJ mol}^{-1}]}{RT}}$$

This is an Arrhenius-type equation.

3. E_a represents the enthalpy of migration $\Delta_m G$ of oxide ions by the vacancy mechanism in solid solutions $(ThO_2)_{1-x}(YO_{1.5})_x$ with $0 < x < 0.01$.

4. **a.** Taking into account the crystalline structure of fluorite, the most likely hop is toward the closest anionic site. The corresponding hopping distance ℓ is a/2, where a denotes the lattice parameter.

   We find $\qquad\qquad\qquad \ell = 2.7975 \text{ Å}$

   b. When the activated hopping model applies, we can write

   $$u_0 = \frac{2v_0 F\ell^2}{RT}$$

   where v_0 is the lattice-vibration frequency.
   We deduce

   $$v_0 = \frac{RTu_0}{2F\ell^2} = \frac{8.314 \times 1149 \times 6.47 \times 10^{-5}}{2 \times 96840 \times (2.7975 \times 10^{-10})^2}$$

$$\nu_0 = 4.1 \times 10^{13} \, s^{-1}$$

This value is compatible with the hopping frequencies in solids.

Solution 3.3 – Study of electronic conductivity in solid solutions $(CeO_2)_{1-x}(CaO)_x$

1. a. We use the following relation to determine the electronic conductivity:

$$\sigma_t = \sigma_i + \sigma_e$$

where σ_t, σ_i, and σ_e denote the total conductivity, ionic conductivity, and electronic conductivity, respectively. We assume that the variation in ionic conductivity with oxygen partial pressure is negligible. The results for the conductivities are presented in table 36.

Table 36 – Total conductivity and electronic conductivity in solid solution $(CeO_2)_{1-x}(CaO)_x$ with $x = 2.8 \times 10^{-3}$ and for several oxygen partial pressures.

P_{O_2} [bar]	1	0.21	10^{-2}	10^{-3}	5×10^{-4}	10^{-4}	10^{-5}
$\log P_{O_2}$ [bar]	0.00	– 0.68	– 2.00	– 3.00	– 3.30	– 4.00	– 5.00
$10^3 \sigma_t$ [S cm^{-1}]	1.096	1.096	1.38	1.66	1.79	2.26	3.14
$\log \sigma_t$ [S cm^{-1}]	– 2.96	– 2.96	– 2.86	– 2.78	– 2.75	– 2.65	– 2.50
$10^3 \sigma_e$ [S cm^{-1}]	0	0	0.284	0.564	0.694	1.164	2.044
$\log \sigma_e$ [S cm^{-1}]	negligible		– 3.55	– 3.25	– 3.16	– 2.93	– 2.69

The curve for the electronic conductivity of $(CeO_2)_{1-x}(CaO)_x$ with $x = 2.8 \times 10^{-3}$ as a function of oxygen partial pressure is shown in figure 56 in logarithmic coordinates.

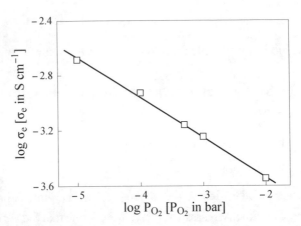

Figure 56 – Electronic conductivity of $(CeO_2)_{1-x}(CaO)_x$ with $x = 2.8 \times 10^{-3}$ as a function of oxygen partial pressure and in logarithmic coordinates.

b. Over the range $10^{-5} < P_{O_2} < 10^{-2}$ bar, we find a linear relationship given by the expression $\log \sigma_e = -0.2857 \log P_{O_2} - 4.1214$

The electronic conductivity is a decreasing function of oxygen partial pressure.

2. Given that the electronic conductivity is a decreasing function of oxygen partial pressure and considering the following equilibrium reaction of the solid solution with gaseous oxygen:

$$O_O^x \rightleftharpoons \tfrac{1}{2}O_{2(g)} + 2e' + V_O^{\bullet\bullet}$$

we note that the electronic conductivity is n type.

3. For an oxygen partial pressure of 10^{-4} bar,

a. the electronic conductivity is

$$\log \sigma_e = -2.98 \ S\,cm^{-1}$$

or
$$\sigma_e = 1.05 \times 10^{-3} \ S\,cm^{-1}$$

b. The expression for ionic transport number is

$$t_i = \frac{\sigma_i}{\sigma_t} = 1 - \frac{\sigma_e}{\sigma_t}$$

Numerical evaluation gives

$$t_i = 1 - \frac{1.05 \times 10^{-3}}{2.26 \times 10^{-3}}$$

$$t_i = 0.535$$

4. a. In the linear range established in question 1, the relation $\sigma_e = f(P_{O_2})$ can be written in the form $\sigma_e = \alpha\, P_{O_2}^{-1/n}$ where α is a constant and $n \approx 3.5$.

b. To interpret this relation theoretically, we must express

▷ the reaction for doping CeO_2 with CaO

$$CaO \longrightarrow Ca''_{Ce} + O_O^X + V_O^{\bullet\bullet}$$

▷ the equilibrium with the gaseous phase

$$O_O^X \rightleftharpoons \tfrac{1}{2} O_2 + V_O^{\bullet\bullet} + 2e'$$

$$K_g = [V_O^{\bullet\bullet}]\, n^2\, P_{O_2}^{\frac{1}{2}}$$

from which we deduce

$$n = K_g^{\frac{1}{2}} \times [V_O^{\bullet\bullet}]^{-\frac{1}{2}} \times P_{O_2}^{-\frac{1}{4}}$$

Assuming that the oxygen vacancies of extrinsic origin dominate ($[V_O^{\bullet\bullet}] = [Ca''_{Ce}]$), we arrive at $n = kP_{O_2}^{-\frac{1}{4}}$ where k is a constant. We find that the equation giving the variation in electronic conductivity with P_{O_2} deduced from this model is close to that established experimentally [see 4(a)].

Solution 3.4 – Electronic transport number in a glass

1. The following electrode reactions occur at

▷ the cathode $\quad\quad Ag^+ + e \longrightarrow Ag$

▷ the anode $\quad\quad Ag \longrightarrow Ag^+ + e$

Considering the activity of Ag^+ ions to be constant in the electrolyte, the emf ΔE at the cell terminals at equilibrium ($i = 0$) is expressed as

$$\Delta E = (1 - \bar{t}_e)\frac{RT}{F} \ln\frac{a_{Ag}^{(2)}}{a_{Ag}^{(1)}}$$

where $\bar{t}_e$ is the average electronic transport number.

2. a. The phenomenon responsible for this variation is the electronic conduction in the electrolyte. Any difference in the silver chemical potential between electrodes (1) and (2) will lead to a transfer of silver between the two compartments. This phenomenon is called electrochemical semipermeability.

b. Becuse $a_{Ag}^{(1)} = 1$, the expression for the emf takes the form

$$\Delta E = (1 - \bar{t}_e)\frac{RT}{F}\ln \gamma_{Ag}^{(2)}x_{Ag}^{(2)}$$

where $\gamma_{Ag}^{(2)}$ and $x_{Ag}^{(2)}$ denote the coefficient of activity and the molar fraction of silver in compartment 2, respectively.

We thus deduce $\quad \dfrac{d\Delta E}{dt} = \dfrac{RT}{F} \times \dfrac{1}{x_{Ag}^{(2)}} \times \dfrac{dx_{Ag}^{(2)}}{dt}$

3. Expression for the electronic conductivity of the electrolyte

By neglecting the overpotentials, the current through the cell may be written as

$$I = \frac{\Delta E}{R_e}$$

where R_e is the electronic resistance of the electrolyte.

The quantity of charge Q traversing the circuit can be expressed in the two forms $\qquad Q = I \times t = n_{Ag}^{(2)} \times F$

where $n_{Ag}^{(2)}$ is the number of moles of silver that traverses the circuit in time t. By denoting by n_T the number of moles (silver + mercury) in compartment 2, we obtain

$$\frac{dQ}{dt} = \frac{\Delta E}{R_e} = \frac{F dn_{Ag}^{(2)}}{dt} = n_T \frac{F dx_{Ag}^{(2)}}{dt}$$

and $\qquad\qquad\qquad dn_{Ag}^{(2)} = n_T \times dx_{Ag}^{(2)}$

We thus deduce

$$\frac{dx_{Ag}^{(2)}}{dt} = \frac{F}{RT} \times x_{Ag}^{(2)} \times \frac{d\Delta E}{dt} = \frac{1}{n_T} \times \frac{\Delta E}{FR_e}$$

Given that the expression relating the electronic conductivity σ_e to the electronic resistance is $\qquad \sigma_e = \dfrac{1}{R_e} \times \dfrac{\ell}{S}$

we obtain $\qquad \sigma_e = \dfrac{F^2}{RT} \times n_T \times x_{Ag}^{(2)} \times \dfrac{\ell}{S} \times \dfrac{1}{\Delta E} \times \dfrac{d\Delta E}{dt}$

4. a. Upper limit to electronic resistance

The expression for the electronic resistance R_e is easily deduced from that established in question 3 for the electronic conductivity:

$$R_e = \frac{RT}{F^2} \times \Delta E \times \frac{1}{n_T} \times \frac{1}{x_{Ag}^{(2)}} \times \frac{1}{\frac{d\Delta E}{dt}}$$

Determine the values for the various factors:

$$T = 300 \text{ K}$$

$$\frac{d\Delta E}{dt} = \frac{10 \times 10^{-3}}{30 \times 24 \times 3600} = 3.858 \times 10^{-9}\ \text{V s}^{-1}$$

$$x_{Ag}^{(2)} = 6 \times 10^{-4}$$

$$n_{Hg} = \frac{m_{Hg}}{M_{Hg}} = \frac{\rho_{Hg} \times V_{Hg}}{M_{Hg}} = \frac{13600 \times 10 \times 10^{-9}}{200 \times 10^{-3}} = 6.8 \times 10^{-4}\ \text{mol}$$

$$x_{Hg} = 1 - x_{Ag} = 1 - 6 \times 10^{-4} = 0.9994$$

$$n_T = \frac{n_{Hg}}{x_{Hg}} = \frac{6.8 \times 10^{-4}}{0.9994} = 6.804 \times 10^{-4}\ \text{mol}$$

$$\Delta E = 0.85\ \text{V}$$

$$\text{or } R_e = \frac{8.314 \times 300}{96\,480^2} \times 0.85 \times \frac{1}{6.804 \times 10^{-4}} \times \frac{1}{6 \times 10^{-4}} \times \frac{1}{3.858 \times 10^{-9}}$$

$$R_e \geq 1.15 \times 10^8\ \Omega$$

b. Upper limit for electronic transport number in glass

Given that
$$\sigma_e = \frac{1}{R_e} \times \frac{\ell}{S}$$

we have
$$\sigma_e \leq \frac{1}{1.45 \times 10^8} \times 0.1$$

$$\sigma_e \leq 6.92 \times 10^{-10}\ \text{S cm}^{-1}$$

Because
$$t_e = \frac{\sigma_e}{\sigma_T} \times \frac{6.92 \times 10^{-10}}{2.2 \times 10^{-3}}$$

we obtain
$$t_e \leq 3.14 \times 10^{-7}$$

Solution 3.5 – Electrical properties of potassium chloride KCl

1. a. The reaction for doping KCl with $BaCl_2$ is

$$BaCl_2 \longrightarrow Ba_K^{\bullet} + 2Cl_{Cl}^{\times} + V_K'$$

The reaction corresponding to the dominant disorder is

$$0 \rightleftharpoons V_K' + V_{Cl}^{\bullet} \qquad \text{with } K_s = [V_K'][V_{Cl}^{\bullet}]$$

where K_s is the Schottky equilibrium constant. Note that the introduction of $BaCl_2$ is accompanied by an increase in the concentration of potassium vacancies and a decrease in the concentration of chlorine vacancies.

b. These variations in concentration are accompanied by an increase in the cationic conductivity and, consequently, by the cationic transport number t_+.

c. The concentration of potassium in its normal position is

$$[K_K^\times] = \frac{\rho}{M}$$

where ρ and M denote the density and molar mass of KCl, respectively. Inserting the corresponding values gives

$$[K_K^\times] = \frac{\rho}{M} = \frac{1.984}{74.56}$$

$$[K_K^\times] = 2.66 \times 10^{-2} \, \text{mol cm}^{-3}$$

Given that the concentration of potassium vacancies is equal to that of barium, we obtain $\quad [V_K'] = x \times [K_K^\times]$

Numerical evaluation gives

$$[V_K'] = 1.4 \times 10^{-4} \times 2.66 \times 10^{-2}$$

$$[V_K'] = 3.72 \times 10^{-6} \, \text{mol cm}^{-3} = 3.72 \times 10^{-3} \, \text{mol L}^{-1}$$

2. At 440 °C (T = 713 K and 1/T [K] = 1.4×10^{-3}), we obtain from figure 44 $\log \sigma_K = -5.5$, so the conductivity $\sigma_K = 3.16 \times 10^{-6} \, \text{S cm}^{-1}$.

Given that $\qquad \sigma_K = F^2 \times \tilde{u}_V \times [V_K']$

the electrochemical mobility $\tilde{u}_V$ of a potassium vacancy is obtained from

the relation $\qquad \tilde{u}_V = \dfrac{\sigma_K}{F^2 [V_K']}$

with $\qquad \tilde{u} = \dfrac{u}{zF}$

Numerical evaluation gives

$$\tilde{u}_V = \frac{3.16 \times 10^{-6}}{96\,480^2 \times 3.72 \times 10^{-6}}$$

$$\tilde{u}_V = 9.12 \times 10^{-11} \, \text{J}^{-1} \, \text{s}^{-1} \, \text{mol cm}^2$$

Using this result in the Nernst-Einstein relation gives

$$D_V = \tilde{u}_V \, RT$$

$$D_V = 9.12 \times 10^{-11} \times 8.314 \times 713$$

$$D_V = 5.4 \times 10^{-7} \, \text{cm}^2 \, \text{s}^{-1}$$

3. a. At 440 °C the diffusion coefficient for the K^+ cation is

$$D_K^* = 2 \times 10^{-5} e^{-\frac{0.76 \times 1.6 \times 10^{-19} \times 6.02 \times 10^{23}}{8.314 \times 713}}$$

or $\qquad D_K^* = 8.66 \times 10^{-11} \text{ cm}^2 \text{ s}^{-1}$

b. We obtain

for K^+ $\qquad D_K^* \times [K_K^x] = 8.66 \times 10^{-11} \times 2.66 \times 10^{-2}$

$$D_K^* \times [K_K^x] = 2.3 \times 10^{-12}$$

and for $[V_K']$ $\quad D_V \times [V_K'] = 5.4 \times 10^{-7} \times 3.72 \times 10^{-6}$

$$D_V \times [V_K'] = 2 \times 10^{-12}$$

These results show that the relation $D_K^* \times [K_K^x] = D_V \times [V_K']$ is approximately verified.

c. The expression for the electrical conductivity is

$$\sigma_K = F^2 \times [V_K'] \times \frac{D_V}{RT} = F^2 \times [K_K^x] \times \frac{D_K^*}{RT}$$

By noting that $[K_K^x]$ and $[V_K']$ are both constant in the extrinsic domain, we deduce that the activation energy of the conductivity (D_V) equals that of the diffusion of potassium (*via* D_K^*).

d. The enthalpy of migration $\Delta_m H_V$ of V_K' is the activation energy evoked in 3(c). It is equal to $\quad \Delta_m H_V = 0.76 \text{ eV}$

or $\qquad \Delta_m H_V = 73.2 \text{ kJ mol}^{-1}$

4. a. Assuming that the mobility of potassium vacancies is independent of the doping level, their concentration in a pure crystal (intrinsic defects) may be calculated from the relation

$$[V_K'] = \frac{D_K^* \times [K_K^x]}{D_V}$$

$$[V_K'] = \frac{4 \times 10^{-2} e^{-\frac{1.48 \times 1.6 \times 10^{-19} \times 6.02 \times 10^{23}}{8.314 \times 713}}}{5.4 \times 10^{-7}} \times 2.66 \times 10^{-2}$$

$$[V_K'] = 7.1 \times 10^{-8} \text{ mol cm}^{-3}$$

b. We deduce the cationic conductivity σ_K by applying the relation

$$\sigma_K = F^2 \times \tilde{u}_V \times [V_K']$$

$$\sigma_K = 96480^2 \times 9.12 \times 10^{-11} \times 7 \times 10^{-8}$$

$$\sigma_K = 5.94 \times 10^{-8}\,\mathrm{S\,cm^{-1}}$$

c. Given that the Schottky equilibrium is

$$0 \rightleftharpoons V_K' + V_{Cl}^{\bullet}$$

we can write for a pure crystal

$$[V_K'] = [V_{Cl}^{\bullet}] = C_0 e^{-\frac{\Delta_f H_S}{2RT}}$$

where $\Delta_f H_S$ is the molar enthalpy of formation of a Schottky pair and C_0 is a constant. The expression for the cationic conductivity is thus

$$\sigma_K = F^2 \times \frac{D_V^{\circ}}{RT} \times e^{-\frac{\Delta_m H_V}{RT}} \times C_0 e^{-\frac{\Delta_f H_S}{2RT}}$$

$$\sigma_K = F^2 \times \frac{D_V^{\circ}}{RT} \times C_0 e^{-\frac{\Delta_m H_V + (\Delta_f H_S/2)}{RT}}$$

By identification, we have

$$E_{a,K} = \Delta_m H_V + (\Delta_f H_S/2)$$

where $E_{a,K}$ is the activation energy of cationic conduction (diffusion) in a pure crystal.

We thus deduce $\quad \Delta_f H_S = 2(E_{a,K} - \Delta_m H_V)$

$$\Delta_f H_S = 2(1.48 - 0.76)$$

$$\Delta_f H_S = 1.44\,\mathrm{eV} = 138.7\,\mathrm{kJ\,mol^{-1}}$$

5. a. Exploiting the relation $D_{Cl}^{*}\,[Cl_{Cl}^{\times}] = D_V\,[V_{Cl}']$ applied to chlorine allows us to express the mobility of a chlorine vacancy, $\tilde{u}_{V,Cl}$, by the relation

$$\tilde{u}_{V,Cl} = \frac{D_{Cl}^{*}}{RT} \times \frac{[Cl_{Cl}^{\times}]}{[V_{Cl}']}$$

Based on the results of the preceding questions and neglecting the effect of doping, we have

$$[V_{Cl}^{\bullet}] = [V_K'] = 7 \times 10^{-8}\,\mathrm{mol\,cm^{-3}}$$

and $\quad [Cl_{Cl}^{\times}] = [K_K^{\times}] = 2.66 \times 10^{-2}\,\mathrm{mol\,cm^{-3}}$

and $\tilde{u}_{V,Cl} = \dfrac{1}{8.314 \times 713} \times \dfrac{2.66 \times 10^{-2}}{7 \times 10^{-8}} \times 200\,e^{-\frac{2.24 \times 1.6 \times 10^{-19} \times 6.02 \times 10^{23}}{8.314 \times 713}}$

$$\tilde{u}_{V,Cl} = 2 \times 10^{-12}\,\mathrm{J^{-1}\,s^{-1}\,mol\,cm^2}$$

Assuming that the activation energy $E_{a,Cl}$ for chlorine self-diffusion is the same as that for anionic conduction in pure KCl, we can write

$$E_{a,Cl} = \Delta_m H_{V,Cl} + (\Delta_f H_S/2)$$

where $\Delta_m H_{V,Cl}$ represents the enthalpy of migration of the chlorine vacancy.

We deduce $\qquad \Delta_m H_{V,Cl} = 2.24 - 1.44/2$

$$\Delta_m H_{V,Cl} = 1.52 \, eV = 146.4 \, kJ \, mol^{-1}$$

Note – We observe that it is significantly greater than that of a potassium vacancy.

b. The anionic transport number is a measure of the contribution of Cl^- ions to the total conductivity of solid KCl. It is given by the ratio

$$t_- = \frac{\sigma_{Cl}}{\sigma_{Cl} + \sigma_K}$$

where σ_{Cl} and σ_K denote the conductivities of Cl^- anions and K^+ cations, respectively.

We know that, at 440 °C,

$$\sigma_K = 5.94 \times 10^{-8} \, S \, cm^{-1} \qquad\qquad \text{(see 4(b))}$$

Moreover, the expression for the anionic conductivity is

$$\sigma_{Cl} = F^2 \times \tilde{u}_{V,Cl} \times [V_{Cl}^{\bullet}]$$

By using the values found above, we obtain

$$\sigma_{Cl} = 96\,480^2 \times 2 \times 10^{-12} \times 7 \times 10^{-8} = 1.3 \times 10^{-9} \, S \, cm^{-1}$$

We thus deduce $\qquad\qquad t_- = 0.021$

c. We write the anionic transport number

$$t_- = \frac{\sigma_{Cl}}{\sigma_{Cl} + \sigma_K}$$

in the form

$$t_- = \frac{F^2 \times \dfrac{D_{Cl}^{*\circ}}{RT} e^{-\frac{E^-}{RT}} \times [Cl_{Cl}^{\times}]}{\left(F^2 \times \dfrac{D_{Cl}^{*\circ}}{RT} e^{-\frac{E^-}{RT}} \times [Cl_{Cl}^{\times}]\right) + \left(F^2 \times \dfrac{D_K^{*\circ}}{RT} e^{-\frac{E^+}{RT}} \times [K_K^{\times}]\right)}$$

which we simplify by using $[Cl_{Cl}^{\times}] = [K_K^{\times}]$

$$t_- = \dfrac{\dfrac{D^{*\circ}_{Cl}}{RT} e^{-\frac{E^-}{RT}}}{D^{*\circ}_{Cl} e^{-\frac{E^-}{RT}} + D^{*\circ}_{K} e^{-\frac{E^+}{RT}}}$$

Solving this relation leads to

$$\ln\left(\frac{1-t_-}{t_-}\, \frac{D^{*\circ}_{Cl}}{D^{*\circ}_{K}}\right) = \frac{E^- - E^+}{RT}$$

By using $t_- = 0.5$, we obtain

$$\ln\frac{D^{*\circ}_{Cl}}{D^{*\circ}_{K}} = \frac{E^- - E^+}{RT} \qquad \text{and} \qquad T = \frac{E^- - E^+}{RT} \times \frac{1}{\ln\dfrac{D^{*\circ}_{Cl}}{D^{*\circ}_{K}}}$$

Numerical evaluation gives

$$T = \frac{(2.24 - 1.48)\times 1.6 \times 10^{-19} \times 6.02 \times 10^{23}}{8.314} \times \frac{1}{\ln\dfrac{200}{4 \times 10^{-2}}}$$

$$T = 1\,033.8 \text{ K}$$

$$T = 760.6\,^{\circ}\text{C}$$

Note that this temperature is close to the melting temperature of KCl. Under these conditions, the anionic conductivity is not negligible compared with the cationic conductivity.

6. The effect of doping KCl with LiCl depends on the position occupied by the lithium.

▷ Li in substitution
The corresponding reaction is

$$\text{LiCl} \longrightarrow \text{Li}^{\times}_{K} + \text{Cl}^{\times}_{Cl}$$

This indicates that no modification of the structure defects responsible for the conductivity is observed. The conductivity is essentially unaffected.

▷ Li in insertion
This is possible given its small size with respect to potassium. The insertion reaction is

$$\text{LiCl} \longrightarrow \text{Li}^{\bullet}_{i} + \text{Cl}^{\times}_{Cl} + \text{V}'_{K}$$

with an increase in the concentration of potassium vacancies and, consequently, in the cationic conductivity.

Solution 3.6 – Application of Nernst-Einstein relation to LiCF$_3$SO$_3$ in poly(ethylene oxide) P(EO)

1. By using the Nernst-Einstein relation that links the electrochemical mobility of lithium, $\tilde{u}_{Li}$, to the diffusion coefficient of lithium, D_{Li}, the ionic conductivity due to lithium may be expressed as

$$\sigma_{Li} = F^2 C_{Li} \tilde{u}_{Li} = \frac{F^2 D_{Li} C_{Li}}{RT}$$

This relation allows us to express the concentration C_{Li} of lithium ions that participate in cationic transport:

$$C_{Li} = \frac{RT\sigma_{Li}}{F^2 D_{Li}}$$

The cationic conductivity σ_{Li} is given by

$$\sigma_{Li} = t_{Li} \times \sigma_t$$

where t_{Li} and σ_t denote the cationic transport number and the total conductivity, respectively.

$$C_{Li} = \frac{RT \times t_{Li} \times \sigma_t}{F^2 D_{Li}}$$

$$C_{Li} = \frac{8.314 \times 358 \times 0.45 \times 9 \times 10^{-5}}{96\,480^2 \times 1.53 \times 10^{-6}}$$

$$C_{Li} = 8.46 \times 10^{-6} \, \text{mol cm}^{-3}$$

The number n_{Li} of effective carriers is thus

$$n_{Li} = C_{Li} \times V$$

where V denotes the sample volume, which is given by

$$V = n \times V_m$$

where n and V_m denote the total number of moles of dissolved TFSI and the molar volume of electrolyte, respectively. The calculation for n gives

$$n = \frac{m}{M} = \frac{48.33}{(2 \times 12 + 4 + 16) \times 36 + 156}$$

$$n = 2.78 \times 10^{-2} \, \text{mol}$$

The molar volume V_m is given by

$$V_m = \frac{M_m}{\rho}$$

M_m and ρ are the molar mass and the density of the polymer electrolyte, respectively.

We obtain $\qquad V_m = \dfrac{1740}{1.3} = 1338.5\,cm^3\,mol^{-1}$

and $\qquad V = 2.78 \times 10^{-2} \times 1\,338.5 = 37.2\,cm^3$

Finally, the number of lithium ions effectively participating in the electric conduction is $\qquad n_{Li} = 8.46 \times 10^{-6} \times 37.2$

$$n_{Li} = 3.15 \times 10^{-4}\,mol$$

2. Assuming that all the free Li^+ ions effectively participate in the electric conduction, the dissociation rate τ for the salt $LiCF_3SO_3$ by the reaction

$$LiCF_3SO_3 \rightleftharpoons Li^+ + CF_3SO_3^-$$

is $\qquad\qquad\qquad \tau = \dfrac{n_{Li}}{n}$

$$\tau = \dfrac{3.15 \times 10^{-4}}{2.78 \times 10^{-2}}$$

$$\tau = 1.13\%$$

Solution 3.7 – Electrical conductivity as a function of composition in $(CeO_2)_{1-x}(YO_{1.5})_x$

1. The curve for the function $\sigma_e = f(x)$ is given in logarithmic coordinates in figure 57.

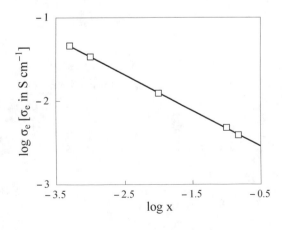

Figure 57 – Electrical conductivity of solid solution $(CeO_2)_{1-x}(YO_{1.5})_x$ as a function of doping rate x in logarithmic coordinates.

2. An analysis of the data leads to the following experimentally based equation:
$$\log \sigma_e = -0.429 \log x - 2.75$$

3. The doping reaction $YO_{1.5} \longrightarrow Y'_{Ce} + \frac{3}{2}O_O^\times + \frac{1}{2}V_O^{\bullet\bullet}$

allows us to write $\qquad [Y'_{Ce}] = 2[V_O^{\bullet\bullet}]$
by neglecting the oxygen vacancies of intrinsic origin.
The equilibrium between the solid solution and the oxygen environment is expressed as

$$O_O^\times \rightleftharpoons \frac{1}{2}O_2 + 2e' + V_O^{\bullet\bullet}$$

with $\qquad K_g = n^2[V_O^{\bullet\bullet}] P_{O_2}^{\frac{1}{2}} \qquad$ and $\qquad \dfrac{[Y'_{Ce}]}{2} = [V_O^{\bullet\bullet}] = \alpha x$

where α is a constant of proportionality. We obtain

$$n = K_g^{\frac{1}{2}}[V_O^{\bullet\bullet}]^{-\frac{1}{2}} P_{O_2}^{-\frac{1}{4}}$$

$$n = K_g^{\frac{1}{2}}\alpha^{-\frac{1}{2}} P_{O_2}^{-\frac{1}{4}} x^{-\frac{1}{2}}$$

At constant temperature and oxygen partial pressure, we have

$$n = Ax^{-\frac{1}{2}}$$

where A is a constant. The expression for the electronic conductivity

$$\sigma_e = u_e F n$$

where u_e is the electric mobility thus takes the form

$$\sigma_e = Bx^{-\frac{1}{2}}$$

where B is a constant.
We obtain the theoretical equation for the electronic conductivity as a function of doping level x

$$\log \sigma_e = -\frac{1}{2}\log x + \log B$$

A comparison with the experimentally determined equation reveals similar slopes, which validates the proposed model.

Solution 3.8 – Conductivity of nickel oxide

1. The conductivity of NiO as a function of oxygen partial pressure at 1 000 °C is shown in figure 58 in logarithmic coordinates.

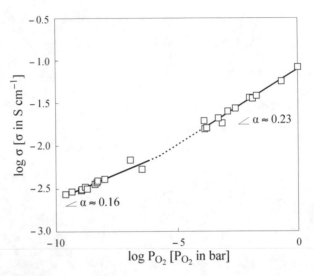

Figure 58 – Conductivity of NiO as a function of oxygen
partial pressure in logarithmic coordinates at 1 000 °C.

We observe a linear conductivity as a function oxygen partial pressure. Two domains with different slopes are apparent. The linear parts of each domain have slopes of 0.16 and 0.23 for lower and higher oxygen partial pressure, respectively.

2. At high oxygen partial pressure, nickel oxide is a p-type semiconductor. It has nickel vacancies and holes in the cationic sublattice and a full anionic sublattice. The reaction with oxygen leads to the formation of atomic and electronic defects

$$\tfrac{1}{2}O_2 \;\rightleftharpoons\; O_O^\times + V_{Ni}'' + 2h^\bullet$$

The equilibrium constant K_{g1} for this reaction is given by

$$K_{g1} = \frac{[V_{Ni}''] \times [h^\bullet]^2}{P_{O_2}^{1/2}}$$

At high oxygen partial pressures, the electroneutrality equation reduces to

$$[h^\bullet] = 2[V_{Ni}'']$$

from which we have

$$K_{g1} = \frac{[h^\bullet]^3}{2P_{O_2}^{1/2}} \quad \text{and} \quad [h^\bullet] = (2 \times K_{g1})^{1/3} \times P_{O_2}^{1/6}$$

If we take the holes to be the majority charge carriers, the total conductivity of the sample is proportional to the hole concentration: $\sigma \propto P_{O_2}^{1/6}$

where $$\log \sigma \propto 0.167 \times \log P_{O_2}$$

This equation holds well enough at low pressure $(10^{-10} < P_{O_2} < 10^{-5})$, with an experimentally determined slope of 0.16 (see fig. 58).

In NiO, holes can be trapped by nickel vacancies *via* the reaction

$$V_{Ni}'' + h^\bullet \rightleftharpoons V_{Ni}'$$

leading to the formation of singly ionized vacancies. The equilibrium reaction with oxygen is thus

$$\tfrac{1}{2}O_{2(g)} \rightleftharpoons O_O^X + V_{Ni}' + h^\bullet$$

The corresponding equilibrium constant K_{g2} is

$$K_{g2} = \frac{[V_{Ni}'] \times [h^\bullet]}{P_{O_2}^{1/2}}$$

Considering the domain where V_{Ni}' dominates, the electroneutrality relation reduces to $$[h^\bullet] = [V_{Ni}']$$

from which we have

$$K_{g2} = \frac{[h^\bullet]^2}{P_{O_2}^{1/2}} \quad \text{and} \quad [h^\bullet] = K_{g2}^{1/2} \times P_{O_2}^{1/4}$$

Assuming that p-type conductivity dominates, we obtain

$$\sigma \propto P_{O_2}^{1/4}$$

or again $$\log \sigma \propto 0.25 \times \log P_{O_2}$$

Experimentally, we find an approximate slope of 0.23 in the domain with high oxygen pressure $(10^{-5} < P_{O_2} < 1)$.

3. The different behavior observed as a function of oxygen partial pressure can be explained by the shift of the equilibrium

$$V_{Ni}'' + h^\bullet \rightleftharpoons V_{Ni}'$$

where the constant K_3 is $$K_3 = \frac{[V_{Ni}']}{[V_{Ni}''] \times [h^\bullet]}$$

with $$K_3 = \frac{K_{g2}}{K_{g1}}$$

This equilibrium is shifted to the left (right) at low (high) oxygen pressure. Because the atomic defects dominate at the highest P_{O_2}, the nickel vacancies are singly ionized.

4. Given the preceding results and considering the slopes found in figure 58, the ionization equilibrium of the nickel vacancies,

$$V''_{Ni} + h^{\bullet} \rightleftarrows V'_{Ni}$$

is shifted to the left at high temperature.

Solution 3.9 – Ionic conductivity-activity relationship of oxide modifier in oxide-based glasses

1. In glasses of the type $(SiO_2)_{1-\alpha}(Na_2O)_\alpha$, silicon dioxide SiO_2 (silica) is the glass former and the monoxide Na_2O is the glass modifier.
   Examples of other glass formers: P_2O_5, B_2O_3.
   Examples of other glass modifiers: K_2O, CaO.

2. In terms of the metal-oxygen bonds in vitreous oxides, the covalence is greater for the glass former. The ionicity is greater for the glass modifier.

3. In silica-based glasses, the coordination index is 4 for silicon (tetrahedral environment) and 2 for oxygen.

4. **a.** Expression for the potential difference ΔE between the terminals of the cell

$$(Hg, Na)^{(1)} / (SiO_2)_{1-\alpha}(Na_2O)_\alpha / (Hg,Na)^{(2)}$$

The following equilibrium holds at each electrode:

$$Na^+ + e \rightleftarrows Na$$

to which corresponds the following electrode potential:

$$E = E^\circ_{Na^+/Na} + \frac{RT}{F} \ln \frac{a_{Na^+}}{a_{Na}}$$

where a_{Na^+} and a_{Na} denote the activities of Na^+ in the vitreous electrolyte and of Na in the amalgam, respectively. From this we derive the expression for the potential difference ΔE between the terminals of the cell:

$$\Delta E = E^{(1)} - E^{(2)}$$

$$\Delta E = \frac{RT}{F} \ln \frac{a^{(1)}_{Na^+}}{a^{(1)}_{Na}} \times \frac{a^{(2)}_{Na}}{a^{(2)}_{Na^+}}$$

Assuming that the Na^+ activity is constant in the vitreous electrolyte and that the Na activity in the mixture is equal to its molar fraction, we obtain

$$\Delta E = \frac{RT}{F} \ln \frac{x_{Na}^{(2)}}{x_{Na}^{(1)}}$$

b. Numerical evaluation gives

$$\Delta E = \frac{8.314 \times 473}{96\,480} \ln \frac{10^{-5}}{2 \times 10^{-3}}$$

$$\Delta E = -216\,\text{mV}$$

5. When the potential difference between the cell terminals goes to zero, the sodium activity in the mixture is the same in compartments 1 and 2. Under these conditions we can write

$$\Delta x_{Na} = x_{Na}^{(1)} - x_{Na}^{f} = x_{Na}^{(2)} + x_{Na}^{f}$$

where x_{Na}^{f} denotes the molar fraction of sodium in each mixture when ΔE is zero.

From this we deduce

$$x_{Na}^{f} = \frac{x_{Na}^{(1)} - x_{Na}^{(2)}}{2}$$

$$x_{Na}^{f} = \frac{2 \times 10^{-3} - 10^{-5}}{2} = 9.95 \times 10^{-4}$$

and

$$\Delta x_{Na} = x_{Na}^{(1)} - x_{Na}^{f}$$

$$\Delta x_{Na} = 2 \times 10^{-3} - 9.95 \times 10^{-4}$$

$$\Delta x_{Na} = 1.005 \times 10^{-3}$$

We now show that the number Δn_{Na} of Na moles transfered from compartment 1 to compartment 2 is given by

$$\Delta n_{Na} = 10^{-2} \Delta x_{Na}$$

Using

$$Q = I \times t = \Delta n_{Na} \times F$$

to denote the electric charge that passes through the cell, we obtain

$$I = \frac{10^{-2} \Delta x_{Na} F}{t}$$

or

$$I = \frac{10^{-2} \times 1.005 \times 10^{-3} \times 96\,480}{60}$$

$$I = 16.2\,\text{mA}$$

6. Crystallization is accompanied by

▷ the appearance of a structural order, and

▷ the formation of microcrystallites separated by grain boundaries.

These two factors manifest themselves in the impedance diagram by an increase in the bulk resistivity of the material and by the appearance of a supplementary resistance at the grain boundaries.

Solution 3.10 – Electrochemical coloration

1. Literal expressions for partial ionic conductivity

For the ionic conductivity σ_i, we use the following expression derived from the Nernst-Einstein relation:

$$\sigma_i = z_i^2 F^2 \frac{D_i}{RT}[i]$$

where [i] denotes the concentration of the ionic species i. The equilibrium involving the Schottky disorder

$$0 \rightleftharpoons V_K' + V_{Br}^{\bullet}$$

allows us to write $K_s = [V_K'][V_{Br}^{\bullet}]$

By neglecting the concentrations of the electronic species (n and p), we have

$$[V_K'] = [V_{Br}^{\bullet}] = \sqrt{K_S}$$

or $$\sigma_K = \frac{F^2}{RT} D_K^{\circ} \sqrt{K^{\circ}} \, e^{-\frac{1.26\,eV}{kT}} e^{-\frac{1.1\,eV}{kT}}$$

$$\sigma_K = \frac{F^2}{RT} D_K^{\circ} \sqrt{K^{\circ}} \, e^{-\frac{2.36\,eV}{kT}}$$

with $$D_K^{\circ} = 10^{-2}\,cm^2\,s^{-1}$$

The same reasoning leads to the expression for the ionic conductivity of bromine:

$$\sigma_{Br} = \frac{F^2}{RT} D_{Br}^{\circ} \sqrt{K^{\circ}} \, e^{-\frac{3.71\,eV}{kT}}$$

2. Assume that the electronic transport number is negligible. The ratio of partial ionic conductivities is

$$\frac{\sigma_K}{\sigma_{Br}} = \frac{D_K^{\circ}}{D_{Br}^{\circ}} e^{-\frac{-2.36+3.71}{kT}} = \frac{D_K^{\circ}}{D_{Br}^{\circ}} e^{-\frac{1.35\,eV}{kT}}$$

Numerical evaluation gives

$$\frac{\sigma_K}{\sigma_{Br}} = \frac{10^{-2}}{3 \times 10^4} e^{-\frac{1.35 \times 1.6 \times 10^{-19}}{1.38 \times 10^{-23} \times 900}}$$

$$\frac{\sigma_K}{\sigma_{Br}} = 11.9$$

Assuming that the electronic transport number is negligible ($t_e \approx t_h \approx 0$), the expression for the transport number for K^+ is

$$t_K = \frac{\sigma_K}{\sigma_K + \sigma_{Br}} = \frac{\frac{\sigma_K}{\sigma_{Br}}}{\frac{\sigma_K}{\sigma_{Br}} + 1}$$

$$t_K = \frac{11.9}{12.9}$$

$$t_K = 0.922$$

3. At the bromine partial pressure $P_{Br_2} = 10^{-10}$ bar, we can read from the diagram

$$[V'_K] = [V^\bullet_{Br}] = 10^{18} \text{ cm}^{-3} \quad \text{and} \quad [h^\bullet] = 10^{14} \text{ cm}^{-3}$$

We thus deduce

$$\sigma_h \approx 10^{-4}\sigma_K$$

and

$$t_h = \frac{\sigma_h}{\sigma_h + \sigma_K + \sigma_{Br}}$$

$$t_h = \frac{1}{1 + 10^4 + 0.084 \times 10^4}$$

$$t_h = 9.22 \times 10^{-5}$$

that we round off to 10^{-4}, which is the order of magnitude of the hole transport number.

4. **a.** The electrochemical chain in question is

$$P_{Br_2}, Me \, / \, KBr_{(s)} \, / \, Me, P_{Br_2}$$
$$\text{(c)} \qquad \qquad \text{(a)}$$

where Me denotes the electrode material, (c) is the cathode, and (a) is the anode.

Each interface is at equilibrium:

$$\tfrac{1}{2} Br_{2(g)} + V^\bullet_{Br(SE)} + e'_{(Me)} \rightleftharpoons Br^\times_{Br}$$

where we verify the following relation:

$$\tfrac{1}{2}\mu^\circ_{Br_2} + \frac{RT}{2} \ln P_{Br_2} + \tilde{\mu}_{V^\bullet_{Br}} + \mu_e - F\varphi = \text{const.}$$

The application to each interface gives

$$\frac{RT}{2} \ln P_{Br_2}^{(c)} + \mu_e^{(c)} - F\varphi^{(c)} = \frac{RT}{2} \ln P_{Br_2}^{(a)} + \mu_e^{(a)} - F\varphi^{(a)}$$

$$\Delta E = \varphi^{(c)} - \varphi^{(a)} = \frac{RT}{2F} \ln \frac{P_{Br_2}^{(c)}}{P_{Br_2}^{(a)}}$$

Because the potential difference $\Delta E = -1$ V,

we have $\ln \dfrac{P_{Br_2}^{(c)}}{P_{Br_2}^{(a)}} = -\dfrac{1 \times 2F}{RT} = -\dfrac{2 \times 96\,480}{8.314 \times 900} = -25.79$

$$P_{Br_2}^{(c)} = 6.3 \times 10^{-12} P_{Br_2}^{(a)}$$

Numerical evaluation gives

$$P_{Br_2}^{(c)} = 6.3 \times 10^{-22} \text{ bar}$$

b. When the electrochemical chain is traversed by a current, the reactions are

▷ at the anode

$$Br_{Br}^{\times} \longrightarrow \tfrac{1}{2} Br_{2(g)} + V_{Br(SE)}^{\bullet} + e_{(Me)}'$$

▷ and at the cathode

$$e_{(Me)}' + V_{Br(SE)}^{\bullet} \longrightarrow V_{Br}^{\times}$$

where $V_{Br}^{\times}$ is an F-type color center and is responsible for the blue coloration of KBr near the cathode.

Note – Before the color appears, the reaction

$$\tfrac{1}{2} Br_{2(g)} + V_{Br(SE)}^{\bullet} + e_{(Me)}' \longrightarrow Br_{Br}^{\times}$$

is what rapidly becomes the limiting factor given the very low bromine partial pressure at the cathode.

c. When the current is stopped, the bromine partial pressure $P_{Br_2}'^{(c)}$ at the cathode can be calculated by the relation established in 4(a)

$$\Delta E = \varphi^{(c)} - \varphi^{(a)} = \frac{RT}{2F} \ln \frac{P_{Br_2}'^{(c)}}{P_{Br_2}^{(a)}}$$

$$P_{Br_2}'^{(c)} = P_{Br_2}^{(a)} \times e^{\frac{2F \times \Delta E}{RT}}$$

$$P'^{(c)}_{Br_2} = 10^{-10} \times e^{\frac{2 \times 96\,480 \times (-2.5)}{8.314 \times 900}}$$

$$P'^{(c)}_{Br_2} = 10^{-38} \text{ bar}$$

Referring to the Brouwer diagram at $P'^{(c)}_{Br_2} = 10^{-38}$ bar , we observe that the concentration of color centers is practically equal to that of the non-ionized bromine vacancies:

$$[V^{\bullet}_{Br(SE)}] = [V^{X}_{Br}] = 10^{18} \text{ cm}^{-3}$$

5. a. The dissociation reaction of solid KBr in the conditions given in the problem statement is

$$KBr_{(s)} \rightleftharpoons K_{(g)} + \frac{1}{2} Br_{2(g)}$$

The corresponding equilibrium constant is

$$K_{eq} = e^{-\frac{\Delta_r G^{\circ}_T}{RT}}$$

$$K_{eq} = e^{-\frac{327\,400}{8.314 \times 900}}$$

$$K_{eq} = 10^{-19}$$

Moreover, we have $\quad K_{eq} = P_K \times P^{\frac{1}{2}}_{Br_2}$
with pressure expressed in bars.
From this we deduce $\quad P_{Br_2} = 10^{-30}$ bar
The electrochemical chain to consider is

$$\underset{(-)}{P_K,Me} / KBr_{(s)} / \underset{(+)}{Me, P_{Br_2}}$$

The crystal becomes colored at the (−) pole, but much less so than for $P_{Br_2} = 10^{-38}$ bar.
The concentrations of the majority species, as read from the diagram (fig. 59), are

$$[V^{X}_{Br}] \approx 10^{15} \text{ cm}^{-3}$$

$$[e'] \approx 10^{9} \text{ cm}^{-3}$$

$$[V^{\bullet}_{Br}] = [V'_K] \approx 10^{18} \text{ cm}^{-3}$$

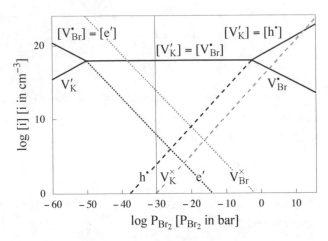

Figure 59 – Brouwer diagram for KBr.

b. A blue-colored front will develop from the (–) pole to the (+) pole.

c. The color front is due to diffusion of electrons, which react with the gaseous bromine whose partial pressure will decrease. Inversely, the species V_K', which is present in excess at the (+) pole, diffuses toward the (–) pole and reacts with the positive gaseous potassium. The net result is the formation of solid KBr (i.e., crystal growth).

d. The potential difference measured between the two poles decreases as the color front progresses.

Note – This behavior is comparable to that of electrochemical semipermeability.

Solution 3.11 – Oxygen diffusion in gadolinia-doped ceria

1. The substitution equation is

$$Gd_2O_3 \longrightarrow 2Gd_{Ce}' + 3O_O^\times + V_O^{\bullet\bullet}$$

The substitution leads to the formation of oxygen vacancies $V_O^{\bullet\bullet}$, which are responsible for the oxide ion mobility in the solid solution.

2. a. The variation of the product DT as a function of temperature in Arrhenius coordinates is shown in figure 60.

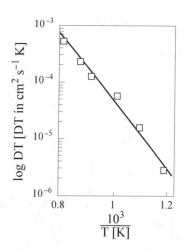

Figure 60 – The product DT as a function of temperature in Arrhenius coordinates.

b. The activation energy is obtained from

$$D = \frac{D_0}{T} e^{-\frac{E_a}{RT}}$$

or

$$\log DT = \log D_0 \left(-\frac{E_a}{2.3\,RT}\right)$$

The linear regression based on the experimental data gives

$$\log(DT) = -5.864 \times \frac{10^3}{T} + 1.567$$

which allows us to calculate $E_a = -19.12 \times (-5.864)$

We obtain $\qquad E_a = 112\,\text{kJ mol}^{-1}$

The activation energy in eV is obtained as follows:

$$E_a[\text{eV}] = E_a[\text{kJ mol}^{-1}] \times \frac{10^3}{e \times N_{Av}}$$

which gives $\qquad E_a = 1.16\,\text{eV}$

3. The electric mobility is given by $u_i = \sigma_i/(c_i z_i e)$. The corresponding dimensions are

$$u_i = \frac{\text{S cm}^{-1}}{\text{cm}^{-3}\,\text{C}}$$

Because $\qquad\qquad C \equiv As \equiv VSs$

we arrive at $\qquad\qquad u_i = \frac{\text{cm}^2}{V\,s}$

The electric mobility of a species i is thus expressed in $\text{cm}^2\,\text{V}^{-1}\,\text{s}^{-1}$.

4. The diffusion coefficient at 800 °C is calculated from the relation

$$\log(DT) = -5.864 \times \frac{10^3}{T} + 1.567$$

from which we have $D_{800\,°C} = \frac{1}{T} \times 10^{\left(-\frac{5864}{T} + 1.567\right)}$

Numerical evaluation gives

$$D_{800\,°C} = \frac{1}{1073} \times 10^{\left(-\frac{5864}{1073} + 1.567\right)}$$

or $D_{800\,°C} = 1.18 \times 10^{-7}\,cm^2\,s^{-1}$

The electric mobility u_i is obtained from the Nernst-Einstein relation

$$u_i = \frac{qD}{kT}$$

where k is the Boltzman constant.

$$u_i = \frac{2 \times 1.6 \times 10^{-19} \times 1.18 \times 10^{-7}}{1.38 \times 10^{-23} \times 1073}$$

We obtain $u_i = 2.55 \times 10^{-6}\,cm^2\,V^{-1}\,s^{-1}$

The solid solution $Ce_{0.7}Gd_{0.3}O_{2-\delta}$ is an ionic conducting material by the vacancy mechanism. Its conductivity σ depends on the concentration of oxygen vacancies

$$\sigma = 2\frac{F}{N_{Av}}[V_O^{··}] \times u$$

The concentration of oxygen vacancies corresponds to the concentration of extrinsic defects obtained by doping with Gd_2O_3. The doping equation and the doping level indicate a quantity of 1.5 vacancies per formula unit of $Ce_{0.7}Gd_{0.3}O_{2-\delta}$. If we consider that half of the vacancies are mobile, we obtain a volume concentration of

$$[V_O^{··}] = 0.5 \times 2 \times 0.15 \times \frac{Z}{a^3}$$

where Z is the number of unit formulas per unit cell. Here, Z = 4.

$$[V_O^{··}] = \frac{4 \times 0.15}{(5.426 \times 10^{-8})^3}$$

We obtain $[V_O^{··}] = 3.756 \times 10^{21}\,cm^{-3}$

and $\sigma = (3.756 \times 10^{21})(2.55 \times 10^{-6})(2 \times 1.6 \times 10^{-19})$

which gives $\sigma = 3.06 \times 10^{-3}\,S\,cm^{-1}$

Solution 3.12 – Electrical conductivity of solid vitreous solution $(SiO_2)_{1-x}(Na_2O)_x$

1. The junction potential between vitreous phases $\langle 1 \rangle$ and $\langle 2 \rangle$ is given by the general expression

$$\varphi_{\langle 1 \rangle} - \varphi_{\langle 2 \rangle} = \frac{1}{F} \int_1^2 \sum_i \frac{t_i}{z_i} d\mu_i$$

Applying this to the given problem leads to

$$\varphi_{\langle 1 \rangle} - \varphi_{\langle 2 \rangle} = \frac{1}{F} \int_1^2 d\mu_{M^+} = \frac{1}{F} \left(\mu_{M^+}^{(2)} - \mu_{M^+}^{(1)} \right)$$

2. Expression of emf ΔE at the terminals of the electrochemical chain

   Each electrode is at the following equilibrium:

   $$\tfrac{1}{2} O_{2(gas)} + 2e_{\langle Pt \rangle} \rightleftharpoons O^{2-}_{\langle glass \rangle}$$

   which allows us to write

   ▷ for electrode 1

   $$\frac{1}{2}\left[\mu_{O_2}^\circ + RT \ln P_{O_2}^{(1)}\right] + 2\left[\mu_e^\circ - F\varphi_{\langle \alpha \rangle}\right] = \mu_{O^{2-}}^{\langle 1 \rangle} - 2F\varphi_{\langle 1 \rangle}$$

   ▷ for electrode 2

   $$\frac{1}{2}\left[\mu_{O_2}^\circ + RT \ln P_{O_2}^{(2)}\right] + 2\left[\mu_e^\circ - F\varphi_{\langle \alpha' \rangle}\right] = \mu_{O^{2-}}^{\langle 2 \rangle} - 2F\varphi_{\langle 2 \rangle}$$

   Given that the oxygen partial pressures $P_{O_2}^{(1)}$ and $P_{O_2}^{(2)}$ are equal and that we have the same electrode metal (Pt), the emf ΔE reduces to

   $$\Delta E = \varphi_{(\alpha)} - \varphi_{(\alpha')}$$

   $$\Delta E = \frac{1}{2F}\left[\mu_{O^{2-}}^{(2)} - \mu_{O^{2-}}^{(1)}\right] + \left[\varphi_{\langle 1 \rangle} - \varphi_{\langle 2 \rangle}\right]$$

   By using the result of question 1, we obtain

   $$\Delta E = \frac{1}{2F}\left[\mu_{O^{2-}}^{(2)} - \mu_{O^{2-}}^{(1)}\right] + \frac{1}{2F}\left[2\mu_{M^+}^{(2)} - 2\mu_{M^+}^{(1)}\right]$$

   By considering that the equilibrium

   $$M_2O \rightleftharpoons 2M^+ + O^{2-}$$

   is realized in the vitreous phases $\langle 1 \rangle$ and $\langle 2 \rangle$, we obtain

   $$\Delta E = \frac{RT}{2F} \ln \frac{a_{M_2O}^{\langle 2 \rangle}}{a_{M_2O}^{\langle 1 \rangle}}$$

3. In the "weak electrolyte" model applied to the amorphous conductors by M^+ cations, the activity of the oxide modifier M_2O is proportional to the square of the charge-carrier concentration, $(a_{M_2O} \propto [M^+]^2)$. If we accept that the mobility is constant, then the square of the conductivity becomes proportional to the activity of the modifier, $(a_{M_2O} \propto \sigma^2)$, which gives

$$\Delta E = \frac{RT}{F} \ln \frac{\sigma^{\langle 2 \rangle}}{\sigma^{\langle 1 \rangle}}$$

4. By using the expression for the ionic conductivity, we arrive at

$$\Delta E = \frac{RT}{F} \ln \frac{\sigma_0^{\langle 2 \rangle}}{\sigma_0^{\langle 1 \rangle}} - \frac{E_a^{\langle 2 \rangle} - E_a^{\langle 1 \rangle}}{F}$$

at $T = 0$, we have $\quad \Delta E_{T=0} = \dfrac{E_a^{\langle 1 \rangle} - E_a^{\langle 2 \rangle}}{F} = \dfrac{\Delta E_a}{F}$

$$\Delta E_a = 0.1 \times 96\,480$$

$$\Delta E_a = 9.650 \text{ kJ mol}^{-1}$$

$$\Delta E_a = 0.1 \text{ eV}$$

Solution 3.13 – High-temperature protonic conductor SrZrO$_3$

A – Point defects in the presence of oxygen

1. The internal equilibria in $SrZrO_3$ are
   ▷ the Schottky equilibrium: $0 \rightleftarrows V''_{Sr} + V^{4\prime}_{Zr} + 3V^{\bullet\bullet}_O$
   with $K_s = [V''_{Sr}]\,[V^{4\prime}_{Zr}]\,[V^{\bullet\bullet}_O]^3$
   ▷ the electronic equilibrium: $0 \rightleftarrows e' + h^\bullet$
   with $K_e = np$

2. The equilibrium for the reaction involving the oxygen partial pressure is

$$O_{2(g)} + 2V^{\bullet\bullet}_O \rightleftarrows 2O^X_O + 4h^\bullet$$

with $K_g = \dfrac{p^4}{[V^{\bullet\bullet}_O]^2 P_{O_2}}$.

B – Point defects in the presence of water vapor

1. The equilibrium reaction of $SrZrO_3$ with water vapor is

$$2H_2O_{(g)} + 2V_O^{\cdot\cdot} \rightleftharpoons 2O_O^X + 4H_i^{\cdot}$$

with $K_{H_2O} = \dfrac{[H_i^{\cdot}]^4}{[V_O^{\cdot\cdot}]^2 P_{H_2O}^2}$, where P_{H_2O} denotes the water partial pressure.

From this we deduce the concentration $[H_i^{\cdot}]$:

$$[H_i^{\cdot}] = K_{H_2O}^{\frac{1}{4}} [V_O^{\cdot\cdot}]^{\frac{1}{2}} P_{H_2O}^{\frac{1}{2}}$$

2. Let us write that the sum of the hydrogen and water vapor partial pressures is constant near the sample:

$$P_{H_2O} + P_{H_2} = \alpha$$

From this we obtain $\quad \dfrac{P_{H_2O}}{\alpha} = \left(1 + \dfrac{P_{H_2}}{P_{H_2O}}\right)^{-1}$

By inserting this result into the expression for $[H_i^{\cdot}]$ from question B.1, we obtain

$$[H_i^{\cdot}] = \dfrac{K[V_O^{\cdot\cdot}]^{\frac{1}{2}}}{\left(1 + \dfrac{P_{H_2}}{P_{H_2O}}\right)^2}$$

with $K = K_{H_2O}^{\frac{1}{4}} \alpha^{\frac{1}{2}}$

3. The equilibrium constant K_{eq} for the formation of water vapor,

$$2H_2 + O_2 \rightleftharpoons 2H_2O$$

is expressed as

$$K_{eq} = e^{-\frac{\Delta_r G_T^0}{RT}}$$

and

$$K_{eq} = e^{-\frac{-494 + 0.112\,T}{8.314 \times 10^{-3}\,T}}$$

$$K_{eq} = \dfrac{P_{H_2O}^2}{P_{H_2}^2 P_{O_2}}$$

from which we obtain $\quad \dfrac{P_{H_2}}{P_{H_2O}} = K_{eq}^{-\frac{1}{2}} P_{O_2}^{-\frac{1}{2}}$

Equation for $[H_i^{\cdot}]$ in terms of oxygen partial pressure (see fig. 47)

Starting from the expression

$$[H_i^{\cdot}] = \dfrac{K[V_O^{\cdot\cdot}]^{\frac{1}{2}}}{\left(1 + \dfrac{P_{H_2}}{P_{H_2O}}\right)^{\frac{1}{2}}}$$

and considering

▷ the high-P_{O_2} range where $P_{H_2}/P_{H_2O} \ll 1$ and $[V_O^{\bullet\bullet}]$ is constant

we can write $\qquad [H_i^{\bullet}] \approx K[V_O^{\bullet\bullet}]^{\frac{1}{2}} = \text{const.}$

▷ the intermediate-P_{O_2} range where $P_{H_2}/P_{H_2O} \gg 1$ and $[V_O^{\bullet\bullet}]$ is constant
in these conditions, the expression for $[H_i^{\bullet}]$ reduces to

$$[H_i^{\bullet}] \approx \frac{K[V_O^{\bullet\bullet}]^{\frac{1}{2}}}{\left(\dfrac{P_{H_2}}{P_{H_2O}}\right)^{\frac{1}{2}}} = KK_{eq}^{\frac{1}{4}}[V_O^{\bullet\bullet}]^{\frac{1}{2}}P_{O_2}^{\frac{1}{4}} = \text{const.} \times P_{O_2}^{\frac{1}{4}}$$

▷ the low-P_{O_2} range where $P_{H_2}/P_{H_2O} \gg 1$ and $[V_O^{\bullet\bullet}]$ increases as P_{O_2} decreases.

4. **a.** Because yttrium substitutes for zirconium, to obtain the solid solution $SrZr_{1-x}Y_xO_{3-0.5x}$, the reaction for doping $SrZrO_3$ by Y_2O_3 is

$$(1-x)SrZrO_3 + 0.5xY_2O_3 + xSrO \longrightarrow Sr_{Sr}^{\times} + (1-x)Zr_{Zr}^{\times} + xY_{Zr}'$$
$$+ (3-0.5x)O_O^{\times} + 0.5xV_O^{\bullet\bullet}$$

The doping leads to an increase in $[V_O^{\bullet\bullet}]$. Under these conditions, the electroneutrality equation (assuming water vapor is present) is

$$n + [Y_{Zr}'] + 2[V_{Sr}''] + 4[V_{Zr}^{4\prime}] = [H_i^{\bullet}] + 2[V_O^{\bullet\bullet}] + p$$

b. At a high level of Y_2O_3 doping, the extrinsic defects dominate. We can simplify the electroneutrality equation as follows:

$$[Y_{Zr}'] + 2[V_{Sr}''] + 4[V_{Zr}^{4\prime}] = [H_i^{\bullet}] + 2[V_O^{\bullet\bullet}]$$

with $\qquad [Y_{Zr}'] = 2[V_O^{\bullet\bullet}] = \text{const.}$

In addition, we have $\qquad [V_{Sr}''] = [V_{Zr}^{4\prime}]$

which gives $\qquad [V_{Sr}''] = \dfrac{K_S^{\frac{1}{2}}}{[V_O^{\bullet\bullet}]^{\frac{3}{2}}} = \text{const.}$

Under these conditions, $[Y_{Zr}']$, $[V_{Sr}'']$, $[V_{Zr}^{4\prime}]$, and $[V_O^{\bullet\bullet}]$ are constant, independent of the oxygen partial pressure.

c. We thus deduce $\quad [H_i^{\bullet}] = 6[V_{Sr}'']$

$$[H_i^{\bullet}] = \frac{6K_S^{\frac{1}{2}}}{[V_O^{\bullet\bullet}]^{\frac{3}{2}}} = \text{const.}$$

A range therefore exists where the protonic conduction is constant independent of the oxygen partial pressure.

C – Determination electronic conduction part

1. a. Theoretical expression for emf ΔE as a function of hydrogen partial pressure $P_{H_2,c}$ and of temperature

Both electrodes are at the following equilibrium:

$$H_{2(g)} \rightleftharpoons 2H_i^{\bullet} + 2e'$$

For electrode k, applying the Nernst equation gives

$$E_k = E_k^{\circ} + \frac{RT}{2F} \ln \frac{[H_i^{\bullet}]_k^2}{P_{H_2,k}}$$

Assuming that the activity of species $H_i^{\bullet}$ is the same at the two electrode-electrolyte interfaces, the expression for the emf ΔE at the cell terminals is

$$\Delta E = \frac{RT}{2F} \ln \frac{P_{H_2,1}}{P_{H_2,c}}$$

or

$$\Delta E = -\frac{RT}{2F} \ln P_{H_2,c}$$

with $P_{H_2,c}$ expressed in bar.

b. This expression is valid under the following conditions:
▷ the electronic conductivity of the electrolyte is negligible,
▷ the protonic conductivity is non-negligible.

c. Table 37 compares the theoretical emf ΔE_{th} with the experimental emf ΔE_{expt} as a function of hydrogen partial pressure $P_{H_2,c}$ at 800 °C. Under these conditions, the expression for the theoretical emf is given by the relation

$$\Delta E_{th} = -\frac{8.314 \times 1073}{2 \times 96\,480} \ln P_{H_2,c}$$

Table 38 compares ΔE_{th} and ΔE_{expt} as a function of temperature for $P_{H_2,c} = 10^{-2}$ bar.

Table 37 – Theoretical emf compared with experimental emf for a cell for several hydrogen partial pressures at 800 °C.

$P_{H_2,c}$ [bar]	6×10^{-2}	2×10^{-2}	10^{-2}
ΔE_{th} [mV]	130	181	213
ΔE_{expt} [mV]	131	183	214

Table 38 – Theoretical emf compared with experimental emf for a cell for several temperatures and with a hydrogen partial pressure of 10^{-2} bar.

T [°C]	600	800	1 000
ΔE_{th} [mV]	173	213	253
ΔE_{expt} [mV]	177	214	254

The good agreement between ΔE_{th} and ΔE_{expt} (within 1% on average) allows us to conclude that the electronic conductivity is negligible and that SrZrO$_3$ doped with Y$_2$O$_3$ may be used as a hydrogen sensor under the temperature and hydrogen partial pressure conditions given in the problem statement.

2. **a.** The flux density J_{H_2} of gaseous H$_2$ is related to the total current density i through the cell by the expression

$$2FJ_{H_2} = t_{H^+} i$$

where t_{H^+} is the protonic transport number. Figure 48 shows that the function $J_{H_2}(i_{H^+})$ is linear. The slope of the line gives the protonic transport number t_{H^+}. We see that the slope does not depend on the composition of the electrolyte used in the cell. The proton transport number is obtained from the slope $\Delta J_{H_2}/\Delta i$ by applying the following relation:

$$t_{H^+} = 2F \times \frac{\Delta J_{H_2}}{\Delta i} \times \frac{1}{V_m}$$

where V_m is the molar volume under STP conditions.

b. Numerical evaluation gives

$$t_{H_i^+} = 2 \times 96\,480 \times 1.15 \times 10^{-7} \times \frac{1}{22.4 \times 10^{-3}}$$

$$t_{H_i^+} = 0.99$$

We can conclude that, under the conditions described in the problem statement, hydrogenated SrZr$_{1-x}$Y$_x$O$_{3-\alpha}$ behaves as a solid proton conducting electrolyte.

3. Measuring the flux of electrochemical semipermeability is an example of a more precise method to determine the proton transport number.

Solution 3.14 – Free-volume model

1. The following hypotheses are put forward in developing the free-volume model for amorphous electrolytes (glasses and polymers):

   ▷ On a molecular scale, a volume V may be associated with each structural unit (atom or mobile segment of a chain). This volume represents the envelope of the displacements of the geometric center of the structural unit, and it increases with temperature.

   ▷ When this volume reaches and exceeds a critical volume V_c, at a temperature T_0, it becomes useful to define the notion of free volume V_f by the relation

$$V_f = V - V_c$$

   V_f is accessible to all mobile species and may be redistributed within an amorphous solid without requiring an external energy source.

   ▷ When the cage occupied by the structural unit attains a free volume greater than a minimal volume V_f^*, the displacement of this unit becomes possible.

   ▷ At a given temperature T and without requiring an external energy source, the free volume oscillates about an average volume $\overline{V}_f$.

2. **a.** The hopping probability for cations is that for which the free volume of the cage occupied by the cation is greater than V_f^*. This probability is given by

$$P = e^{-\frac{V_f^*}{\overline{V}_f}}$$

   b. Under these conditions, the hopping probability as a function of temperature is

$$P = e^{-\frac{V_f^*}{V_0\,\Delta\alpha(T-T_0)}}$$

   c. By identification, we have

$$P = e^{-\frac{V_f^*}{V_0\,\Delta\alpha(T-T_0)}} = e^{-\frac{B}{R(T-T_0)}}$$

   which gives
$$B = \frac{R\,V_f^*}{V_0\,\Delta\alpha}$$

   B has the same dimensions as $R(T - T_0)$; namely, energy per mole. It is not an activation energy in the classic sense because the hop requires no thermal energy.

3. Study of conductivity of the system $2SiO_2-K_2O$

  a. ▷ Figure 61 shows the curve for the function $\log \sigma_+ T = f(\frac{1}{T})$. The curve is linear from 124 °C to 501 °C. At higher temperature ($T \geq 501$ °C), the experimental points depart from linearity.

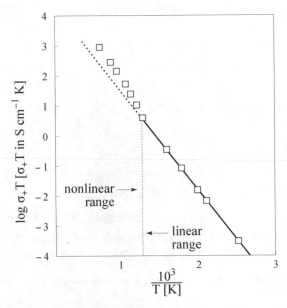

Figure 61 – The function $\log \sigma_+ T = f(\frac{1}{T})$.

By considering the Arrhenius equation $\sigma_+ T = A_1 e^{-\frac{E_a}{RT}}$, a linear regression allows us to determine the constants A_1 and E_a for the low-temperature range.

We obtain $\qquad E_a = 63\ 677 \ J\ mol^{-1}$

and $\qquad\qquad A_1 = 7.59 \times 10^4 \ S\ cm^{-1}K$

The final equation is $\quad \sigma_+ T = 7.59 \times 10^4 e^{-\frac{63\,677}{RT}}$

  ▷ E_a is the activation energy of the cationic conduction in the vitreous system $2SiO_2-K_2O$. It is related to the enthalpy of formation and of migration of the charge carrier by the following expression:

$$E_a = \frac{\Delta_f H}{2} + \Delta_m H$$

  b. ▷ The empirical relations linking the ideal glass transition temperature T_0, the glass transition temperature T_G, and the melting temperature T_m, allow us to estimate T_0: $T_0 = \frac{3}{4} T_G$

$$T_0 = \frac{3}{4} \times 757$$

$$T_0 = 568 \, \text{K}$$

A least squares refinement of the parameters A_2 and B_2 in the non-linear range gives the best fit between the experimental points

and theory $\qquad \sigma_+ T = A_2 e^{-\frac{B_2}{R(T-T_0)}}$

- The parameters obtained for the high-temperature part are

$$A_2 = 4.17 \times 10^3 \, \text{S cm}^{-1} \, \text{K}$$

$$B_2 = 12.34 \, \text{kJ mol}^{-1}$$

- The equation for the product $\sigma_+ T$ is

$$\sigma_+ T = 4.17 \times 10^3 e^{-\frac{12\,340}{R(T-568)}}$$

- The experimental function $\log \sigma_+ T = f(\frac{1}{T})x$ and the equations corresponding to the linear and non-linear ranges are shown in figure 62. The equations are prolonged into high temperatures for the linear range and into low temperatures for the non-linear range.

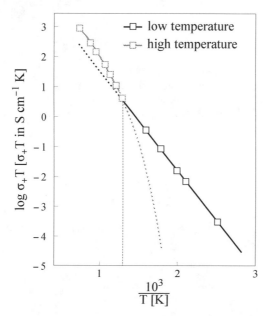

Figure 62 – Conductivity in Arrhenius coordinates. The squares represent the experimental points and the lines show the theory.

We find good agreement between the experimental points and the theoretical models.

▷ The function $\log \sigma_+ T = f\left(\frac{1}{T-T_0}\right)$ in the high temperature range is shown in figure 63. The curve is linear. A linear regression leads to the following equation:

$$\log \sigma_+ T = -0.645 \times \frac{10^3}{T - T_0} + 3.62$$

with $T_0 = 568$ K

From this equation, we obtain values for A_2 and B_2.

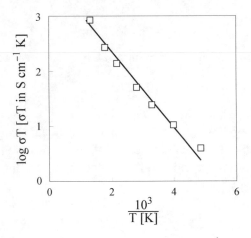

Figure 63 – The function $\log \sigma_+ T = f\left(\frac{1}{T-T_0}\right)$
in the high-temperature range.

▷ The ratio $\frac{V_f^*}{V_0}$ is determined from the relations

$$\frac{B_2}{R} = \frac{V_f^*}{V_0 \, \Delta\alpha}$$

$$\frac{V_f^*}{V_0} = \frac{\Delta\alpha \, B_2}{R}$$

$$\frac{V_f^*}{V_0} = \frac{(1.455 \times 10^{-4} - 5.58 \times 10^{-5}) \times 12\,340}{8.314}$$

$$\frac{V_f^*}{V_0} = 0.13$$

4. Figure 62 indicates that the two ranges intersect at a point near 500 °C. The temperature T_0 is near 300 °C. Thermally assisted hopping contributes starting at the temperature T_0, but overtakes thermally activated hopping (i.e., Arrhenius type) only starting at 500 °C.

Solution 3.15 – Study of single-crystal calcium fluoride CaF$_2$ in the presence of oxygen

1. Equilibria involved

   ▷ intrinsic equilibrium: $F_F^{\times} \rightleftharpoons V_F^{\bullet} + F_i'$

   ▷ introduction of oxygen: $\frac{1}{2}O_{2(g)} + 2F_F^{\times} \rightleftharpoons O_F' + V_F^{\bullet} + F_{2(g)}$

2. The introduction of oxygen, which substitutes for fluorine in the crystal, causes an increase in the concentration of fluorine vacancies, which are of extrinsic origin. This translates into an increase in the ionic conductivity of the crystal by the vacancy mechanism.

3. **a.** Calculation of the conductivity of CaF$_2$ at 661 °C under argon

   Taking into account the quantities given in the problem statement, the expression to calculate the conductivity is

   $$\sigma = \frac{1}{R} \times \frac{\ell}{S} = \frac{1}{R} \times \frac{4\ell}{\pi d^2}$$

   where R, ℓ, and S are the ohmic resistance, the length, and the sample surface area, respectively. The resistance R is deduced by extrapolation to low frequencies of the circular arc in figure 49. We obtain R = 15 kΩ.

   From this, $\quad \sigma = \dfrac{1}{15 \times 10^3} \times \dfrac{0.816 \times 4}{\pi (0.690)^2}$

   $$\sigma = 1.45 \times 10^4 \, S\,cm^{-1}$$

   b. ▷ Activation energy in domains A, B, and C

   We are dealing with a transport phenomenon where migration is at work in each temperature range. The activation energies may be determined from the slopes of the lines in figure 50. The expression for the activation energy E_i obtained from the Arrhenius equation

   $$\sigma = \frac{\sigma_0}{T} e^{-\frac{E_i}{RT}}$$

   is $\quad E_i = 2.3 \times \dfrac{R}{\dfrac{1}{T_2} - \dfrac{1}{T_1}} \times \dfrac{1}{1.6 \times 10^{-19}} \times \dfrac{1}{6.02 \times 10^{23}} \times \log \dfrac{\sigma_1 T_1}{\sigma_2 T_2}$

R is the ideal gas constant, $\sigma_1 T_1, T_1$ and $\sigma_2 T_2, T_2$ are the coordinates of two points 1 and 2 situated within the same linear temperature range. We obtain the following activation energies:

$$E_A = 2.01 \text{ eV}$$

$$E_B = 0.65 \text{ eV}$$

▷ The temperature range with the smallest activation energy (range B) corresponds to pure migration. We thus deduce that the activation energy E_m for migration is

$$E_m = E_B = 0.65 \text{ eV}$$

In this range, the expression for the ionic conductivity as a function of temperature is

$$\sigma = \frac{\sigma_0}{T} e^{-\frac{E_m}{RT}}$$

where σ_0 is a constant. The ionic conductivity is implemented by fluorine vacancies of extrinsic origin (doping with oxygen and impurities).

In range A, the expression for the activation energy is

$$E_A = E_m + \frac{1}{2} E_f$$

where E_f is the formation energy for the Frenkel pair corresponding to the following reaction:

$$F_F^{\times} \rightleftharpoons V_F^{\bullet} + F_i'$$

We thus deduce $E_f = 2.72 \text{ eV}$

In this temperature range, the expression for ionic conductivity as a function of temperature is

$$\sigma = \frac{\sigma_0'}{T} e^{-\frac{E_m + \frac{E_f}{2}}{RT}}$$

c. Conductivity as a function of oxygen partial pressure in range B

At constant temperature, the expression for ionic conductivity in CaF_2 is

$$\sigma = F \times u_{V_F^{\bullet}} \times [V_F^{\bullet}]$$

where $u_{V_F^{\bullet}}$ and $[V_F^{\bullet}]$ denote the electric mobility and the fluorine vacancy concentration, respectively.

The following equilibrium exists:

$$\tfrac{1}{2}O_{2(g)} + 2F_F^\times \;\rightleftharpoons\; O_F' + V_F^\bullet + F_{2(g)}$$

with
$$K_{eq} = \frac{[O_F'][V_F^\bullet]P_{F_2}}{P_{O_2}^{\frac{1}{2}}}$$

By using the hypotheses of
▷ constant fluorine partial pressure at the electrode,
▷ $[V_F^\bullet] = [O_F']$ in the extrinsic range,
we arrive at $\qquad [V_F^\bullet] = A \times P_{O_2}^{\frac{1}{4}}$
where A is a constant. By using this in the expression above for the conductivity, we finally arrive at $\;\sigma \propto P_{O_2}^{\frac{1}{4}}$
This theoretical model takes into account the experimentally observed variation.

Solution 3.16 – emf of membrane crossed by an electrochemical semipermeability flux

1. The equilibrium $\quad \tfrac{1}{2}O_{2(surface)} + 2e_{MO} \;\rightleftharpoons\; O_{MO}^{2-}$

   allows us to write $\qquad \tfrac{1}{2}\mu_{O_2}^* + 2\tilde{\mu}_e^{MO} = \tilde{\mu}_{O^{2-}}^{MO}$

   Because the electrons and the O^{2-} species are in the same phase MO_{1-x}, we have

   $$\tfrac{1}{2}\mu_{O_2}^* + 2\mu_e^{MO} - 2F\varphi^{MO} = \mu_{O^{2-}}^{MO} - 2F\varphi^{MO}$$

   or
   $$\tfrac{1}{2}\mu_{O_2}^* + 2\mu_e^{MO} = \mu_{O^{2-}}^{MO}$$

2. The equilibrium involving the electrons associated with the gold and with the non-stoichiometric compound is $e_{Au} \rightleftharpoons e_{MO}$

   with $\qquad \tilde{\mu}_e^{Au} = \tilde{\mu}_e^{MO}$

3. Because no flux crosses the MO_{1-x}/YSZ interface, the equilibrium

   $$O_{MO}^{2-} \;\rightleftharpoons\; O_{YSZ}^{2-}$$

   is attained, which gives $\qquad \tilde{\mu}_{O^{2-}}^{MO} = \tilde{\mu}_{O^{2-}}^{YSZ}$

4. Figure 64 shows the qualitative variation of the potential in the chain with the potential difference ΔE between the extremities.

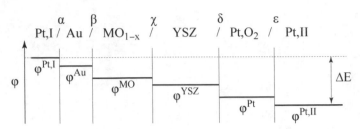

Figure 64 – Qualitative variation in potential within the chain.

5. Because the compound MO_{1-x} is mainly an electronic conductor, oxygen displacement becomes the limiting process. Measuring the flux gives us access to the ionic conductivity in MO_{1-x}.

6. The emf of the chain is expressed as

$$\Delta E = \varphi^{Pt,I} - \varphi^{Pt,II}$$

The equilibria at each interface and the corresponding equations are gathered below.

▷ Electronic equilibrium at interfaces α and β

$$\tilde{\mu}_e^{Pt,I} = \tilde{\mu}_e^{Au} = \tilde{\mu}_e^{MO}$$

▷ Ionic equilibrium at interface χ

$$\tilde{\mu}_{O^{2-}}^{MO} = \tilde{\mu}_{O^{2-}}^{YSZ}$$

▷ Electrochemical equilibrium at interface δ

$$\tfrac{1}{2}O_{2(gas)} + 2e_{Pt} \rightleftharpoons O^{2-}_{YSZ}$$

$$\tfrac{1}{2}\mu_{O_2} + 2\tilde{\mu}_e^{Pt} = \tilde{\mu}_{O^{2-}}^{YSZ}$$

▷ Electronic equilibrium at interface ε

$$e_{Pt} \rightleftharpoons e_{Pt,II}$$

$$\tilde{\mu}_e^{Pt,II} = \tilde{\mu}_e^{Pt}$$

Combining the preceding relations leads to

$$\tilde{\mu}_e^{Pt,I} = \tilde{\mu}_e^{MO} \qquad \text{and} \qquad \tfrac{1}{2}\mu_{O_2} + 2\tilde{\mu}_e^{Pt,II} = \tilde{\mu}_{O^{2-}}^{MO}$$

or

$$2\mu_e^{Pt,I} - 2F\varphi^{Pt,I} = 2\mu_e^{MO} - 2F\varphi^{MO}$$

and $\quad \frac{RT}{2} \ln P_{O_2} + 2\mu_e^{Pt} - 2F\varphi^{Pt,II} = \mu_{O^{2-}}^{MO} - 2F\varphi^{MO}$

By developing the equation given in question 1,

$$\frac{RT}{2} \ln P_{O_2}^* = \mu_{O^{2-}}^{MO} - 2\mu_e^{MO}$$

we obtain $\quad \Delta E = \varphi^{Pt,I} - \varphi^{Pt,II} = \frac{RT}{4F} \ln\left(\frac{P_{O_2}^*}{P_{O_2}}\right)$

Solution 3.17 – Determination of electronic conductivity by electrochemical semipermeability

1. The expression for electrochemical semipermeability flux as a function of partial pressures P_1 and P_2 on each side of the stabilized zirconia membrane is

$$J_{O_2} = \frac{KRTu_h}{4\ell}(P_2^{\frac{1}{4}} - P_1^{\frac{1}{4}})$$

where K is a constant that depends on the equilibrium constant of the reaction

$$\tfrac{1}{2}O_2 + V_O^{\cdot\cdot} \rightleftarrows O_O^{\times} + 2h^{\cdot}$$

u_h is the electric mobility of holes and ℓ is the membrane thickness.
This relation is valid when the electronic conductivity is due mainly to electrons.

2. At equilibrium, the flux of oxygen adsorption at the membrane surface is the same as that for desorption, $J_{ad} = J_{des}$
and the oxygen activity at the membrane surface is the same as that for oxygen in the surrounding gas.
In the presence of an oxygen flux J_{sp} due to the electrochemical semiper-meability of the membrane, we have

$$J_{ad} \pm J_{sp} = J_{des}$$

This appears as a departure from equilibrium at the surface that grows with increasing semipermeability flux or with decreasing partial pressure in the gas. The membrane surface is not in equilibrium with the gas in terms of oxygen activity. Under these conditions, the hypothesis proposed by the Wagner theory is not valid.

3. Figure 65 shows a qualitative diagram of the variation in potential within the cell.

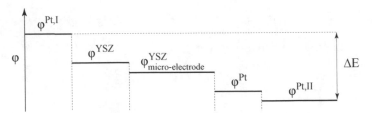

Figure 65 – Qualitative variation in potential within the chain.
ΔE is the potential difference between the chain terminals.

4. The expression for the potential difference ΔE between the external platinum wires I and II is

$$\Delta E = \varphi^{Pt,I} - \varphi^{Pt,II} = \frac{RT}{4F} \ln\left(\frac{P^*_{O_2}}{P_{O_2}}\right)$$

We thus deduce

$$P^*_{O_2} = P_{O_2} e^{\frac{4F\Delta E}{RT}}$$

Numerical evaluation gives

$$P^*_{O_2} = 0.5 \times 10^{-6} e^{\frac{4 \times 96\,480 \times 0.04}{8.314 \times 1725}}$$

$$P^*_{O_2} = 1.47 \times 10^{-6} \text{ bar}$$

The measured emf allows us to estimate the departure from equilibrium at the membrane surface. If a partial pressure is high on one side of the membrane, we can consider that the membrane remains in equilibrium.

5. Based on the experimentally determined P*, we obtain

$$P_3^{\frac{1}{4}} - P^{*\frac{1}{4}} = 0.52 \text{ bar}^{\frac{1}{4}}$$

The semipermeability flux is determined graphically from figure 53

$$J_{O_2} = 53 \times 10^{-9} \text{ mol cm}^{-2} \text{ min}^{-1}$$

or

$$J_{O_2} = 8.8 \times 10^{-10} \text{ mol cm}^{-2} \text{ s}^{-1}$$

6. If P_3 is much greater than P_2, the semipermeability of oxygen as a function of the electronic conductivity of the electrolyte on each side of the membrane is

$$J_{O_2} = \frac{RT}{4F^2 \ell} \sigma_h$$

7. Based on the preceding relation, we have

$$\sigma_h = \frac{4F^2\ell}{RT} J_{O_2}$$

$$\sigma_h = \frac{4 \times 96\,480^2 \times 0.4}{8.314 \times 1725} \times 8.8 \times 10^{-10}$$

$$\sigma_h = 9.1 \times 10^{-4}\,\mathrm{S\,cm^{-1}}$$

Chapter 4

Electrode reactions

Course notes
Thermodynamics and electrochemical kinetics

4.1 – Electrode thermodynamics

4.1.1 – Electrode

An electrode is a system made up of an electronic or mixed conductor (metal or electrode material) in contact with a solid electrolyte and is host to one or several oxidation-reduction reactions.

4.1.2 – Electrode potential

The electrode potential is the electrical potential of an electrode measured with respect to a reference electrode or other comparative electrode. With zero current, it is called the open-circuit potential $E_{i=0}$. When the electrode is host to a single electrochemical reaction in equilibrium,

$$Ox + ne \rightleftharpoons Red$$

the open-circuit potential E_{th} is given by the Nernst equation

$$E_{th} = E^{\circ}_{Ox/Red} + \frac{RT}{nF} \ln \frac{[Ox]_e}{[Red]_e}$$

by associating the activities of the various species to their concentrations $[Ox]_e$ and $[Red]_e$. E_{th} is treated as a thermodynamic or theoretical potential. The standard potential $E^{\circ}_{Ox/Red}$ depends on the two species involved in the reaction.

© Springer Nature Switzerland AG 2020
A. Hammou and S. Georges, *Solid-State Electrochemistry*,
https://doi.org/10.1007/978-3-030-39659-6_4

Thus, we will refer to the O_2/O^{2-} couple for an oxygen electrode and to the Li^+/Li couple for a lithium electrode.

4.1.3 – Electrode polarization

Polarizing an electrode means that you impose on it a potential E that differs from its open-circuit potential. The polarization is given by

$$\Pi = E - E_{i=0}$$

If $\Pi > 0$, the electrode undergoes an oxidation reaction and carries a positive current. If $\Pi < 0$, the electrode undergoes a reduction reaction and, according to convention among electrochemists, carries a negative current.

4.1.4 – Electrode overpotential

The electrode overpotential is the polarization of an electrode subject to a single electrochemical reaction

$$\eta = E - E_{th}$$

4.1.5 – Current density

Current density is the current I per unit surface area S of the electrode-electrolyte interface:

$$i = \frac{I}{S}$$

4.2 – Electrochemical kinetics

4.2.1 – Review

◇ *Steady state and transient state*

In electrochemistry, the system is in a steady state when the potential, the current, and the concentration of electroactive species are independent of time. If this is not the case, the system is in a transient state and the concentrations vary in time according to kinetic equations such as the Fick's law for diffusion and the equations for the heterogeneous kinetics of adsorption (see Appendix).

◇ *Kinetic regimes*

In solid-state electrochemistry, the most common kinetic regimes are the regimes of charge transfer (or activation), diffusion, and adsorption-desorption. We limit ourselves in what follows to expressing the overpotential of an electrode as a function of current density in the electrode and of the kinetic parameters relevant to each regime.

4.2.2 – Pure-charge-transfer regime (extreme case)

This regime appears when the rate constants of diffusion and adsorption of the electroactive species are much higher than the rate constant of electronic transfer.

◇ Exchange-current density

Consider the following electrode reaction at equilibrium:

$$Ox + ne \rightleftarrows Red$$

The electrode is the site of two opposing reactions, each characterized by a current density $i_{0,Red} < 0$ and $i_{0,Ox} > 0$ such that

$$i_{0,Ox} = |i_{0,Red}| = i_0$$

i_0 denotes the exchange-current density, which is expressed in $A cm^{-2}$.

$i_{0,Red}$ and $i_{0,Ox}$ are given by the following expressions:

$$i_{0,Ox} = nFk_{Ox}[Red]_0 e^{\frac{\alpha nF}{RT}E_{th}} \quad and \quad i_{0,Red} = -nFk_{Red}[Ox]_0 e^{-\frac{\beta nF}{RT}E_{th}}$$

where k_{Ox} and k_{Red} are the respective rate constants for the oxidation and reduction reactions and $[Ox]_0$ and $[Red]_0$ are the concentrations of Ox and Red at the electrode, which are constant in their given phase. α and β denote the anodic and cathodic transfer coefficients, respectively, with $\alpha + \beta = 1$.

◇ Butler-Volmer equation

The current density i through a polarized electrode subjected to an overpotential η is the sum of the oxidation current i_{Ox} and the reduction current i_{Red}:

$$i = i_{Ox} + i_{Red}$$

The Butler-Volmer equation expresses i as a function of the overpotential as follows:

$$i = i_0 \left(e^{\frac{\alpha nF}{RT}\eta} - e^{-\frac{\beta nF}{RT}\eta} \right)$$

◇ Electrode impedance Z

The electrode impedance is obtained by expanding the Butler-Volmer equation in a Taylor series to linearize it and then applying a Laplace transform. We denote by ΔX the change in the magnitude of X and by $\overline{X}$ the transform in the Laplace plane. The Butler-Volmer equation thus becomes

$$\overline{\Delta i} = i_0 \left(\frac{\alpha n F}{RT} e^{\frac{\alpha n F}{RT} \eta} \overline{\Delta \eta} + \frac{\beta n F}{RT} e^{-\frac{\beta n F}{RT} \eta} \overline{\Delta \eta} \right)$$

For a given overpotential η, the impedance Z is

$$Z = \frac{\overline{\Delta \eta}}{\overline{\Delta i}} = \frac{1}{i_0} \times \frac{RT}{\alpha n F e^{\frac{\alpha n F}{RT} \eta} + \beta n F e^{-\frac{\beta n F}{RT} \eta}} = R_t$$

This gives the resistance R_t to charge transfer and is generally expressed in $\Omega \, cm^2$. At equilibrium ($\eta = 0$), it is given by

$$R_t^{eq} = \frac{RT}{n F i_0}$$

◇ *Tafel equation*

Under strong anodic polarization, we have $i \approx i_{Ox} = i_0 \, e^{\frac{\alpha n F}{RT} \eta}$ that we normally write in the form $\eta = a + b \ln i$, which is known as the Tafel equation

with
$$a = -\frac{RT}{\alpha n F} \ln i_0 \quad \text{and} \quad b = \frac{RT}{\alpha n F}$$

4.2.3 – Mixed transfer-diffusion regime

When chemical species undergo diffusion, we consider the linearized concentration profiles of the Ox and Red species, shown in figure 66, with the electrode serving as reference ($x = 0$). This schematic assumes that Ox and Red are in the same phase, which is quite infrequent in solid-state electrochemistry. δ_{Ox} and δ_{Red} denote the thickness of the diffusion layers. Under these conditions, the anodic and cathodic current densities are

$$i_{Ox} = n F D_{Red} \frac{[Red]_\varphi - [Red]_e}{\delta_{Red}} \quad \text{and} \quad i_{Red} = -n F D_{Ox} \frac{[Ox]_e - [Ox]_\varphi}{\delta_{Ox}}$$

where D_{Red} and D_{Ox} are the diffusion coefficient of the reducing and the oxidizing agents, respectively.

The index φ denotes the zone where the electroactive species is independent of distance from the electrode surface. The diffusion process may occur in the electrolyte, in a bulk electrode, or in the gas phase containing the electroactive species.

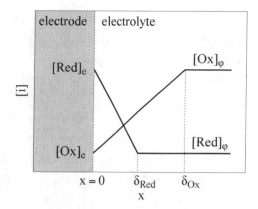

Figure 66 – Linearized concentration profiles of electroactive
species as a function of distance from electrode surface.

◇ Diffusion limit current

If we sufficiently increase the polarization, the current densities will tend to
their limiting values $i_{\ell_{Ox}}$ and $i_{\ell_{Red}}$, where

$$i_{\ell_{Ox}} = \frac{nFD_{Red}}{\delta_{Red}}[Red]_{\varphi} = k_{Red}[Red]_{\varphi}$$

and

$$i_{\ell_{Red}} = -\frac{nFD_{Ox}}{\delta_{Ox}}[Ox]_{\varphi} = -k_{Ox}[Ox]_{\varphi}$$

They correspond to zero concentration at the electrode. k_{Ox} and k_{Red} are the
rate constants associated with the species Ox and Red, respectively.

◇ Overpotential equation in mixed charge-transfer-diffusion regime

This is given by the relation

$$i = i_0 \frac{e^{\frac{\alpha nF}{RT}\eta} - e^{-\frac{\beta nF}{RT}\eta}}{1 + \frac{i_0}{i_{\ell_{Ox}}}e^{\frac{\alpha nF}{RT}\eta} - \frac{i_0}{i_{\ell_{Red}}}e^{-\frac{\beta nF}{RT}\eta}}$$

4.2.4 – Regime of pure diffusion kinetics (extreme case)

This is the case where the charge transfer reaction is much faster than diffusion;
in other words, when the exchange current tends to infinity. We then arrive at
the following overpotential equation:

$$\eta = \frac{RT}{nF}\ln\frac{1 - \frac{i}{i_{\ell_{Red}}}}{1 - \frac{i}{i_{\ell_{Ox}}}}$$

Figure 67 shows the diffusion overpotential curve for an Ox/Red couple. From this curve we can access the concentrations $[Ox]_\varphi$ and $[Red]_\varphi$. This property is exploited in amperometric sensors (see chapter 5). Moreover, we show that the potential measured for $i = (i_{\ell_{Ox}} + i_{\ell_{Red}})/2$ is the half wave potential $E_{1/2}$, which is expressed as

$$E_{1/2} = E^\circ_{Ox/Red} + \frac{RT}{nF} \ln \frac{D_{Red}}{D_{Ox}} \times \frac{\delta_{Ox}}{\delta_{Red}}$$

This potential is characteristic only of the given Ox/Red couple, independent of concentrations $[Ox]_\varphi$ and $[Red]_\varphi$. It allows us to determine whether this couple is present at the working electrode.

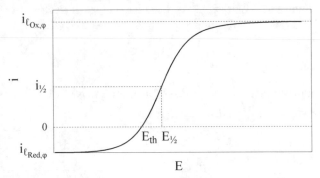

Figure 67 – Polarization curve for a system leading to a reduced interfacial concentration of the electroactive species (diffusion regime).

◇ *Impedance of electrode undergoing diffusion-limited reaction*

▷ *Case of semi-infinite diffusion*

The impedance, which is commonly called the Warburg impedance, is written as

$$Z = \frac{RT}{F^2} \times \frac{\delta_{Ox}}{[Ox]_\varphi \sqrt{2D_{Ox}}} \times \omega^{1/2}(1 - j)$$

▷ *Case of diffusion in a layer of thickness δ*

We treat the case of a reduction reaction. The overpotential equation is

$$\eta = \frac{RT}{nF} \ln\left(1 - \frac{i}{i_{\ell_{Red}}}\right)$$

The impedance is given by the following expression:

$$Z = R_d \times \frac{th\sqrt{j\omega \frac{\delta_{Ox}^2}{D_{Ox}}}}{\sqrt{j\omega \frac{\delta_{Ox}^2}{D_{Ox}}}}$$

where R_d denotes the diffusion resistance, which varies with overpotential. At equilibrium, the expression for R_d is

$$R_d^{eq} = \frac{RT}{n^2 F^2} \times \frac{\delta_{Ox}}{[Ox]_\varphi D_{Ox}}$$

Note – In the case of a gas, $[Ox]_\varphi$ is a function of the partial pressure of the gas.

Figure 68 shows the general form of the impedance diagram in the Nyquist plane for the two cases discussed above.

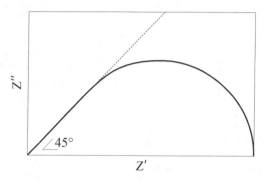

Figure 68 – General form of impedance diagram for a diffusion-limited reduction reaction in a semi-infinite layer (dotted) and for diffusion in a layer of thickness δ (solid curve).

4.2.5 – Regime of adsorption of gaseous species

We consider an electrode reaction controlled by the adsorption of a gaseous species X, taking into account the X_{ads} / X^- couple. The adsorption happens on the electronic conductor (metal or electrode material) or on the solid electrolyte. In these conditions, the charge-transfer reaction should be at equilibrium,

$$X_{ads} + e \rightleftarrows X^-$$

and the electrode overpotential is given by

$$\eta = \frac{RT}{F} \ln \frac{a_{X_{ads}}}{a_{X_{ads},eq}}$$

$a_{X_{ads}}$ and $a_{X_{ads},eq}$ are the respective activities of X_{ads} when polarized and at equilibrium. $a_{X_{ads}}$ depends on the fractional surface coverage θ of the electrode. Limiting ourselves to the case in which the adsorbed phase may be considered a dilute solution ($\theta \approx 0$), we can write $a_{X_{ads}} = \theta$ and

$$\eta = \frac{RT}{F} \ln \frac{\theta}{\theta_{eq}}$$

To express the overpotential, we write the sum of the flux densities

$$\frac{i}{F} + J_{ads} - J_{des} = 0$$

where J_{ads} and J_{des} are respectively the flux density of adsorption and desorption. For a dilute adsorbed phase, we can write $J_{ads} = bP$ and $J_{des} = b'\theta$ where b and b' are constants and $P = P_{X_2}$ for a non-dissociative adsorption and $P = P_{X_2}^{1/2}$ for a dissociative adsorption. In the steady state regime, we arrive at the following expression for the overpotential:

$$\eta = \frac{RT}{F} \ln\left(1 - \frac{i}{i_\ell}\right)$$

i_ℓ is a limiting current given by $i_\ell = -FbP$. Near equilibrium, we have $\eta = (RT/F) \times (i/i_\ell)$ and a polarization resistance R_p with $R_p = RT/(F^2bP)$.

In the periodic regime and at equilibrium, we show that the impedance Z of the electrode is

$$Z = \frac{RT}{F^2bP} \times \frac{1}{(1+j\omega/b')}$$

which we can represent a parallel circuit $R_{ads}//C_{ads}$ with

$$R_{ads}^{eq} = \frac{RT}{F^2} \times \frac{1}{b'\theta_{eq}} = \frac{RT}{F^2} \times \frac{1}{bP} \quad \text{and} \quad C_{ads}^{eq} = \frac{F^2}{RT}\theta_{eq} = \frac{F^2}{RT} \times \frac{bP}{b'}$$

Exercises

Exercise 4.1 – Oxygen-diffusion-limited electrode

We study an oxygen electrode deposited on an O^{2-}-ion conducting electrolyte. In all that follows, we assume that the temperature is such that the only limiting step is the diffusion in the electrode of the oxide ion O^{2-}.

1. Write the reaction mechanism that corresponds to this hypothesis and specify the slow step and the fast steps.

2. To simplify the calculation, we write $a_{O^{2-}}(z) = X$ for the oxygen activity at a distance z from the electrolyte-electrode interface. In this case, the limiting conditions are
   ▷ at $t = 0$, $X = X^0$, where X^0 is the oxygen activity at equilibrium for a given oxygen pressure,
   ▷ at $t > 0$, $z = 0$, in the steady state, $X = X_i$. For very large z (z going to infinity), we write $X = X^0 = $ constant.

   Furthermore, we use the following fundamental relations:
   ▷ Fick's first law $\qquad\qquad J_{O^{2-}} = -D \dfrac{\partial X}{\partial z}$

   ▷ Fick's second law $\qquad \dfrac{\partial X}{\partial t} = D \dfrac{\partial^2 X}{\partial z^2}$

   ▷ nernstian overpotential η $\quad \eta = \dfrac{RT}{2F} \ln \dfrac{X_i}{X^0}$

   and we denote by S the electrode-electrolyte contact surface area.
   a. Draw a schematic representation of X as a function of z at $t = 0$ in the transient state and the steady state in a diffusion layer of thickness δ.
   b. Show that applying Fick's first law leads to
   $$\left(\frac{\partial X}{\partial z} \right)_{z=0} = -\frac{I}{2FSD}$$

3. By using an electric signal I, we impose a small periodic fluctuation on X around X^0 according to $X = X^0 + \Delta X e^{j\omega t}$. By using this expression in Fick's second law,

a. show that we obtain the solution

$$X_i = X^0 + \frac{I}{2FS\sqrt{j\omega D}}$$

b. deduce the expression for the overpotential η.

4. Starting from the derivative of the overpotential with respect to I near the origin ($I \approx 0$), or by doing a polynomial approximation

   a. Show explicitly that the impedance obtained is a Warburg impedance.

   Data

   $$\left| \; j^{-\frac{1}{2}} = \frac{1}{\sqrt{2}}(1-j) \right.$$

   b. For a symmetric cell containing two identical oxygen-diffusion-limited electrodes deposited on each side of a sintered solid, draw the general form of the impedance diagram in both the Nyquist and Bode representations. Represent and explain the different elementary contributions of the response of the electrochemical cell.

   c. How does a point of fixed frequency evolve as a function of oxygen pressure in the gas surrounding the electrode material?

Exercise 4.2 – Study of oxygen-electrode reaction

We use impedance spectroscopy to study the electrochemical behavior of the lanthanum cobaltite $La_{0.7}Sr_{0.3}CoO_{3-\delta}$. Circular electrodes are deposited in a symmetric configuration on an yttria-stabilized zirconia (YSZ) electrode by radiofrequency sputtering. They have a diameter d = 20 mm and a thickness ℓ = 0.35 µm. The electrochemical cell is inserted in a tube furnace. Each electrode is connected to a frequency response analyzer (FRA) *via* platinum meshes and wires. A schematic diagram of the cell appears in figure 69.

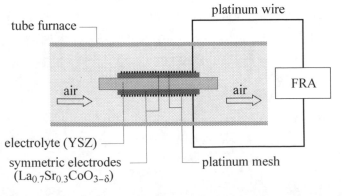

Figure 69 – Schematic diagram of cell and experimental setup.

Figure 70 shows the impedance diagram obtained in air after stabilization at 425 °C in the frequency range of $10^4 - 10^{-3}$ Hz.

This diagram is modeled by the equivalent electric circuit shown in figure 70. For high frequencies (f > 100 Hz), the model consists of an inductance L and the circuit (R // CPE)$_1$. At low frequencies (f < 100 Hz), the model consists of the circuit (R // CPE)$_2$ and at very low frequencies (f < 0.5 Hz) the model consists of a Warburg-type diffusion-limited element whose impedance is

$$Z(\omega) = A \times \frac{\tanh\sqrt{\tau j\omega}}{\sqrt{\tau j\omega}} \tag{1}$$

where ω is the frequency of the electric field, and A and τ are adjustable parameters. The complete expression for Warburg diffusion-limited impedance is

$$Z(\omega) = \frac{RT}{n^2 F^2} \times \frac{\delta}{C^0 D} \times \frac{\tanh\sqrt{j\omega\frac{\delta^2}{D}}}{\sqrt{j\omega\frac{\delta^2}{D}}} \tag{2}$$

where δ is the thickness of the diffusion layer, D is the diffusion coefficient, C^0 is the interfacial concentration of the electroactive species, T is the absolute temperature, and n is the number of electrons exchanged in the electrochemical reaction in question. We consider only the low-frequency range (f < 100 Hz).

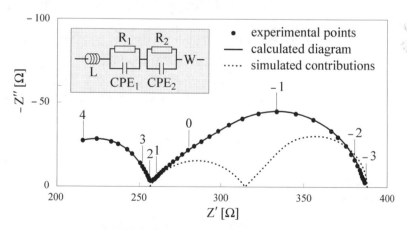

Figure 70 – Impedance diagram obtained under air after stabilization at 425 °C (from Ringuedé & Guindet, 1997).

By a least squares fit of the theoretical model (for the electrode reaction) to the experimental points, we determine the parameters given in table 39.

Table 39 – Parameters for equivalent electric
circuit obtained from fit to experimental points.

W	A [Ω]	τ [s]	(R//CPE)$_2$	R$_2$ [Ω]	CPE$_2$ [F s^{p-1}]	p$_2$
	73.3	1.425		58.1	6.78×10^{-3}	0.63

1. Determine the nature of the electrochemical response observed above and below 100 Hz. How can we ensure that these contributions are properly identified?

2. Based on equations (1) and (2), establish the expression and determine the units for the dc Warburg resistance R_W so as to obtain an interfacial charge-carrier concentration in $mol\,cm^{-3}$ and a diffusion coefficient in $cm^2\,s^{-1}$. Calculate the resistance R_W.

3. By considering that the thickness of the diffusion layer equals the electrode thickness, calculate the diffusion coefficient for oxygen in the electrode and the interfacial concentration of oxide ions.

4. Calculate the area specific resistance (ASR) of this electrode at 425 °C.

Exercise 4.3 – Overpotential in an oxygen electrochemical pump

We consider an oxygen electrochemical pump (fig. 71) consisting of an yttriated zirconia tube of external diameter $\varphi = 2$ cm and thickness $\ell = 2$ mm. Two electrodes, composed of silver paint annealed at 800 °C, are deposited on both side of the tube over a length of 20 cm. The two electrodes are connected to a potentiostat P. The exterior of the pump is in contact with air at a pressure of 1 bar. The gas that circulates inside the pump has an oxygen partial pressure P_e at the entrance of 10^{-3} bar. We want to filter out the oxygen so as to obtain an exit pressure $P_{ex} = 10^{-5}$ bar. The operating temperature of the pump is fixed at 640 °C and it has a flow rate of 5 L h^{-1}.

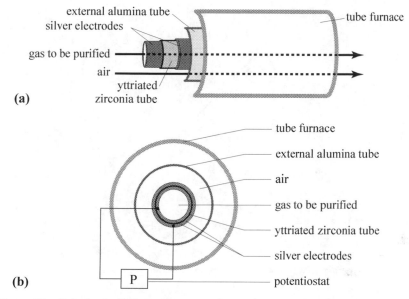

Figure 71 – Schematic diagram of **(a)** experimental apparatus and **(b)** cross section.

The overpotential η_c of the cathode is dictated by the oxygen flux over the silver electrode and is given by

$$\eta_c = \frac{RT}{4F} \ln \left(1 - \left| \frac{I}{I_{\ell c}} \right| \right)$$

where $I_{\ell c}$ is the current limit of the cathode.

Figure 72 shows the normalized cathodic polarization resistance as a function of the oxygen partial pressure.

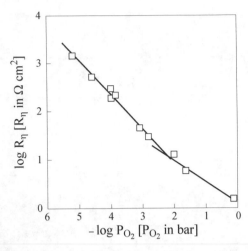

Figure 72 – Normalized cathodic polarization resistance as a
function of oxygen partial pressure in logarithmic coordinates.

1. Given that Schottky disorder is the dominant disorder in pure zirconia, which
   defect gives rise to the ionic conduction in stabilized zirconia?

2. **a.** Derive the expression for the current I required to filter the gas in the
   tube.
   b. Calculate the current.

3. By using the data in figure 72, determine the diffusion limit current $I_{\ell c}$ given
   R_η. For this,
   ▷ assume an oxygen partial pressure given by the arithmetic mean of the
   entrance and exit pressures of the tube and
   ▷ assume $I \ll I_{\ell c}$.

4. Verify that, in this case, we indeed obtain $|I| < |I_{\ell c}|$. What happens if we
   exceed this limit?

5. **a.** Express $I_{\ell c}$ as a function of oxygen partial pressure for the experimental
   conditions inside the tube and in the regime in which it is valid.
   b. What happens to this equation under conditions of high oxygen partial
   pressure; for example, under air?

6. We know that ionic conductivity of the electrolyte in use is $1.78 \times 10^{-2}\,\text{S cm}^{-1}$
   at 727 °C and that its variation with temperature is

$$\sigma = \frac{A}{T} e^{-\frac{0.85\,eV}{kT}}$$

where A is a constant and k is the Boltzmann constant.

a. Calculate the electrolytic resistance.

b. Deduce the emf $|\Delta U|$ generated in this experiment if we ignore the anodic overpotential.

Exercise 4.4 – Determination of exchange current

We want to determine the exchange current I_0 that corresponds to the O_2,Pt / composite / YSZ electrode, where YSZ is the solid electrolyte $(ZrO_2)_{0.92}(Y_2O_3)_{0.08}$ and the composite is a mix of YSZ and the electrode material $La_{0.8}Sr_{0.2}MnO_{3-\delta}$. Figure 73 shows a schematic diagram of the experimental cell.

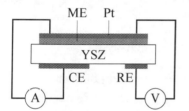

ME: measuring electrode (composite)
CE: counter electrode (Pt)
RE: reference electrode (Pt)
YSZ: yttria stabilized zirconia
A: ammeter
V: voltmeter

Figure 73 – Schematic diagram of three-electrode cell.

The reference electrode (RE) and the counter electrode (CE) are made from platinum. Table 40 gives the results for the current I as a function of overpotential η of the electrode, in air and at 747 °C.

Table 40 – Current as a function of electrode overpotential.

η [mV]/air	0	− 5	− 10	− 15	− 20	− 25	− 30	− 35	− 40	− 47
I [mA]	0	− 0.33	− 0.68	− 1.00	− 1.34	− 1.66	− 2.01	− 2.33	− 2.65	− 3.13
η [mV]/air	–	5	10	15	20	30	40	–	–	–
I [mA]	–	0.33	0.67	1.01	1.33	2	2.66	–	–	–

1. Graph the electrode overpotential as a function of the current I and state its formula.

2. **a.** Write

   ▷ the adsorption-desorption reaction for oxygen dissociation and the corresponding adsorption rate. Denote by Γ the number of adsorption sites per unit surface area, by θ the fractional surface coverage, and by k_{ads} and k_{des} the respective adsorption and desorption constants, assuming that the adsorption follows the Langmuir equation with $\theta \ll 1$.

▷ the equation relating the coverage rate at equilibrium θ_{eq} to the oxygen partial pressure.

▷ the charge-transfer reaction at the triple phase boundary between the gas, the electrode material, and the electrolyte and the corresponding rate. Denote by E the potential applied to the electrode and by α the anodic transfer coefficient. These reactions are generally accepted for modeling the reduction of oxygen at the $(La_{0.8}Sr_{0.2}MnO_{3-\delta}\text{-}YSZ)/$ YSZ interface.

b. Deduce the formula for the current density i flowing through the electrode as a function of potential E.

3. Derive the formula for the exchange-current density i_0.

4. a. Express the formula for the current density under polarization as a function of the exchange-current density and the electrode overpotential η.

b. What happens to this formula when the fractional surface coverage is negligible and the electrode is only weakly polarized?

c. Deduce the formula for the polarization η as a function of the current I and of the contact surface area S.

5. Compare the theoretical formula obtained in question 4(a) to that observed experimentally in question 1 and deduce the exchange current I_0 of the electrode under study.

Exercise 4.5 – Reduction of water vapor at the M / YSZ interface with M = Pt, Ni

Table 41 lists the results of potentiostatic measurements of the current I as a function of the cathodic overpotential η corresponding to the reduction of water vapor at 860 °C at the Pt / YSZ and Ni / YSZ interfaces.

Table 41 – Overpotential and cathodic current at 860 °C.

					At the Pt / YSZ interface							
η [mV]	0	– 25	– 50	– 100	– 150	– 200	– 300	– 400	– 500	– 600	– 700
I [μA]	0	– 9	– 16	– 26.5	– 33.5	– 38	– 43	– 45.2	– 46.3	– 46.7	– 46.8
					At the Ni / YSZ interface							
η [mV]	0	– 25	– 50	– 100	– 150	– 200	– 300	– 400	– 500		
I [μA]	0	– 0.651	– 1.054	– 1.397	– 1.521	– 1.568	– 1.591	– 1.596	– 1.598		

1. Graph the current as a function of the overpotential for each type of electrode.

2. These results are interpreted by a formula for the overpotential of the form

$$\eta = \frac{RT}{zF} \ln \frac{I_\ell - I}{I_\ell}$$

where z is the number of electrons exchanged and I_ℓ is the diffusion limit current.

a. Graphically determine the limited current I_ℓ for each electrode.

b. For each electrode, determine z and propose a possible resulting charge-transfer reaction.

Exercise 4.6 – Hydrogen oxidation at Ni / YSZ interface

We consider hydrogen oxidation at a composite anode (cermet Ni-YSZ) at the Ni / YSZ interface at 975 °C and in the presence of a hydrogen-water mixture with a ratio $P_{H_2}/P_{H_2O} = 15.45$.

Note – YSZ represents yttria-stabilized zirconia.

1. Calculate the oxygen partial pressure under these conditions.

    Data

    for H_2O at 975 °C:
    Standard enthalpy of formation $\Delta_f H^\circ_{1248\,K} = – 224.205 \text{ kJ mol}^{-1}$
    Standard entropy of formation $\Delta_f S^\circ_{1248\,K} = – 45.73 \text{ J K}^{-1} \text{ mol}^{-1}$

2. Table 42 lists the results for anodic overpotential η for this electrode as a function of applied current density i.

Table 42 – Anodic overpotential of electrode as a function of current density.

i [mA cm^{-2}]	2	5	10	20	35	50	85	135	200	300	400
η [V]		0.013	0.029	0.052	0.086	0.120	0.145	0.182	0.217	0.247	0.277	0.299

a. Propose a reaction model that describes the behavior of an anode with two electrons available for the charge-transfer reaction.

b. When the oxidation kinetics is limited by the charge-transfer step, the characteristic current density i and overpotential η are related by the Butler-Volmer equation:

$$i = i_0 \left(e^{\frac{\alpha n F}{RT} \eta} - e^{-\frac{(1-\alpha) n F}{RT} \eta} \right)$$

▷ Determine the exchange-current density i_0 and the anodic transfer coefficient α.

▷ Give the experimental formula for the current density as a function of the overpotential obtained for this electrode.

c. On a single plot, graph the curves $\eta = f(\log i)$ representing the experimental points and the formula obtained in 2(b). Comment.

d. What type of formula can we use for high current densities? Determine this formula.

Solutions to exercises

Solution 4.1 – Oxygen-diffusion-limited electrode

1. The reaction mechanisms to reduce oxygen lead to the following successive steps:
   ▷ diffusion in the gaseous phase and dissociative adsorption at the surface of the electrode material

$$O_{2(g)} + 2s \longrightarrow 2O\text{-}s \qquad\qquad \text{rapid}$$

   where s is an adsorption site.
   ▷ a charge-transfer reaction at the surface of the electrode material

$$O\text{-}s + 2e + V_O^{\cdot\cdot} \longrightarrow O_O^{\times} + s \qquad\qquad \text{rapid}$$

   ▷ diffusion of the species $O_O^{\times}$ (O^{2-}) into the solid electrode $\qquad$ slow
   ▷ exchange at the interface

$$O_O^{\times}\text{ (electrode)} \longrightarrow O_O^{\times}\text{ (electrolyte)} \qquad\qquad \text{rapid}$$

2. **a.** Figure 74 schematically shows X as a function of z at t = 0 and in the transient and steady states.

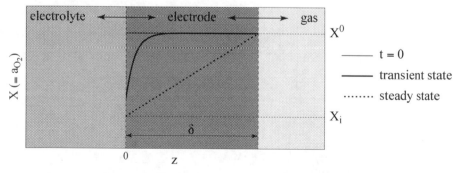

Figure 74 – X as a function of z at t = 0 in a semi-infinite medium in the transient state and in a boundary layer of thickness δ in the steady state.

   b. The relationship between the current density i and the flux density $J_{O^{2-}}$ of O^{2-} ions at the gas/electrode interface in the steady state is

$$J_{O^{2-}} = \frac{i}{2F}$$

or
$$\frac{i}{2F} = -D\left(\frac{\partial X}{\partial z}\right)_{z=0}$$

and
$$\left(\frac{\partial X}{\partial z}\right)_{z=0} = -\frac{I}{2FSD}$$

3. **a.** Given the relation $X = X^0 + \Delta X_{(z)} e^{j\omega t}$, we start by differentiating X with respect to time
$$\frac{\partial X}{\partial t} = j\omega \Delta X_{(z)} e^{j\omega t}$$

$$D\frac{\partial^2 X}{\partial z^2} = D\frac{\partial^2 \Delta X_{(z)}}{\partial z^2} e^{j\omega t}$$

Applying Fick's second law gives
$$\frac{\partial X}{\partial t} = D\frac{\partial^2 X}{\partial z^2}$$

so we obtain
$$D\frac{\partial^2 \Delta X_{(z)}}{\partial z^2} = j\omega \Delta X_{(z)}$$

which we write in the form
$$\frac{\partial^2 \Delta X_{(z)}}{\partial z^2} - \frac{j\omega}{D} \Delta X_{(z)} = 0$$

Integration of this differential equation gives
$$\Delta X_{(z)} = \alpha\, e^{-\sqrt{\frac{j\omega}{D}} \times z} + \beta\, e^{\sqrt{\frac{j\omega}{D}} \times z}$$

We show that $\beta = 0$. For $z = 0$, this gives
$$\Delta X_{(0)} = \alpha$$

Moreover, we have
$$\frac{\partial X}{\partial z} = \frac{\partial \Delta X_{(z)}}{\partial z} e^{j\omega t}$$

and
$$\frac{\partial \Delta X_{(z)}}{\partial z} = -\alpha\sqrt{\frac{j\omega}{D}} e^{-\sqrt{\frac{j\omega}{D}} \times z}$$

which gives, for $z = 0$,
$$\left(\frac{\partial X}{\partial z}\right)_{z=0} = -\alpha\sqrt{\frac{j\omega}{D}} e^{j\omega t} = -\frac{I}{2FSD}$$

By replacing α by $\Delta X_{(z=0)}$, we obtain
$$\Delta X_{(z=0)} e^{j\omega t} = \frac{I}{2FS\sqrt{j\omega D}}$$

Identification with the equation $X = X^0 + \Delta X_{(z)} e^{j\omega t}$ gives
$$X = X^0 + \frac{I}{2FS\sqrt{j\omega D}}$$

b. Given the expressions for the overpotential η and the activity X_i, we can write

$$\eta = \frac{RT}{2F} \ln \left(1 + \frac{I}{2FSX^0 \sqrt{j\omega D}} \right)$$

Near equilibrium (I = 0), the polynomial approximation of $\ln(1+x) = x$ allows us to write

$$\eta = \frac{RT}{2F} \times \frac{I}{2FSX^0 \sqrt{j\omega D}}$$

$$\eta = \frac{RT}{4\sqrt{2}\,F^2 S} \times \frac{1}{X^0} \times \frac{1-j}{\sqrt{\omega D}} \times I$$

4. a. The impedance Z of the electrode is expressed as

$$Z = \frac{RT}{4\sqrt{2}\,F^2 S} \times \frac{1}{X^0} \times \frac{1-j}{\sqrt{\omega D}}$$

which we can write in the form

$$Z = \sigma_W \omega^{-\frac{1}{2}}(1-j)$$

This is a Warburg impedance that varies with $\omega^{-\frac{1}{2}}$ and with a Warburg coefficient of

$$\sigma_W = \frac{RT}{4\sqrt{2}\,F^2 S} \times \frac{1}{X^0} \times \frac{1}{\sqrt{D}}$$

b. Figure 75 shows the form of the impedance diagram of the electrochemical cell in the (a) Nyquist representation and (b) Bode representation.

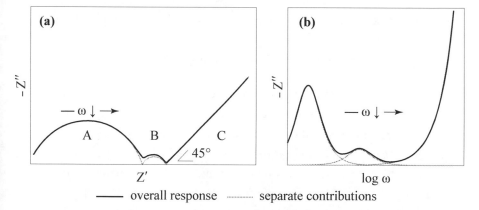

— overall response ┈┈┈ separate contributions

Figure 75 – General form of impedance diagram in the **(a)** Nyquist representation and **(b)** Bode representation of a symmetric cell with two identical electrodes to an oxygen-limited-diffusion electrode deposited on a sintered solid electrolyte.

We can distinguish two major types of response:

▷ Electrical phenomena related to the movement of charge carriers due to an applied ac electric field (domains A and B); for example,
- the bulk response of electrolyte (A), which corresponds to intra-grain ionic migration phenomena. These fast phenomena appear at high frequencies.
- the interface response at grain boundaries (B), which correspond to inter-grain ionic migration phenomena. These phenomena are slower and appear at intermediate frequencies. The influence of grain boundaries leads to a relative increase in the electrical resistance of ceramics with respect to the corresponding single crystals. We thus speak of the "blocking effect" of carriers. These blocking effects depend strongly on the ceramic microstructure.

▷ Electrochemical phenomena involving complex reaction processes such as diffusion, adsorption, and charge transfer (domain C), which together constitute the electrode polarization. These much slower phenomena appear at low frequencies. The form of the impedance diagram depends on limiting phenomena at the electrodes.

In the case of an oxygen-diffusion-limited electrode, the polarization appears in the Nyquist representation in the form of a line with slope ½, which is characteristic of this type of mechanism.

c. If we assume that the oxygen activity X_O is an increasing function of oxygen partial pressure, the expression for impedance Z shows that the impedance at a fixed frequency decreases with oxygen partial pressure.

Solution 4.2 – Study of oxygen-electrode reaction

1. The phenomena observed at high frequencies are related to the properties of the electrolyte. The inductance reflects inductive parasitic effects. At low frequency, we observe the overall reaction at the oxygen electrode. Some of these reactions depend on the thickness and on the surface area of the electrolyte surface, whereas others depend solely on the electrode surface. Varying the geometric factor allows us to differentiate between these reactions. Furthermore, as opposed to the electrolyte resistance, the electrode reaction depends on the oxygen partial pressure. Controlling this pressure thus constitutes another way to identify the electrode reaction.

2. The impedance of a diffusive element in a diffusion boundary layer may be

expressed as
$$Z(\omega) = \frac{RT}{n^2F^2} \times \frac{\delta}{C^0D} \times \frac{\tanh\sqrt{j\omega\frac{\delta^2}{D}}}{\sqrt{j\omega\frac{\delta^2}{D}}}$$

For dc current (i.e., for $\omega \to 0$),
$$Z(\omega) \to \frac{RT}{n^2F^2} \times \frac{\delta}{C^0D}$$

which gives the Warburg resistance, or the diffusion resistance, R_W
$$R_W = \frac{RT}{n^2F^2} \times \frac{\delta}{C^0D}$$

The electrode considered here is a cathode. It is thus the site of the electro-chemical reduction of oxygen

$$\tfrac{1}{2}O_{2(g)} + 2e + V_O^{\ddot{}} \longrightarrow O_O^{\times}$$

We take n = 2. Table 43 lists the units of the relevant quantities.

Table 43 – Units of relevant quantities.

Quantity	C^0	D	R	T	F	d
Units	$mol\,cm^{-3}$	$cm^2\,s^{-1}$	$J\,K^{-1}\,mol^{-1}$	K	$C\,mol^{-1}$	cm

By using
$$1\,J = 1\,\Omega A^2 s$$

and
$$1\,C = 1\,As$$

the equation for the dimensions of R_W becomes

$$\frac{J\,K^{-1}\,mol^{-1}\,K\,cm}{C^2\,mol^{-2}\,mol\,cm^{-3}\,cm^2\,s^{-1}} = \frac{\Omega\,A^2\,s\,K^{-1}\,mol^{-1}\,K\,cm}{A^2\,s^2\,mol^{-2}\,mol\,cm^{-3}\,cm^2\,s^{-1}} = \Omega\,cm^2$$

To obtain an interface concentration in $mol\,cm^{-3}$, we must express resistance in $\Omega\,cm^2$.

The measured electrode resistance is 73.3 Ω. Because the sample is symmetric, the resistance of one electrode is half (i.e., $\frac{A}{2}$), which gives 36.65 Ω. The active surface area of the electrode is

$$S = \pi\left(\frac{d}{2}\right)^2$$

and
$$R_W = \frac{A}{2}S$$

$$R_W = 36.65 \times \pi$$

We find $$R_W = 115\,\Omega\,\text{cm}^2$$

3. For a least squares fit, we use the following form:

$$Z(\omega) = R_W \times \frac{\tanh\sqrt{\tau j \omega}}{\sqrt{\tau j \omega}}$$

where R_W and τ are fitting parameters.

By identification, we obtain $$D = \frac{\delta^2}{\tau}$$

$$D = \frac{(0.35 \times 10^{-4})^2}{1.425}$$

$$D = 8.6 \times 10^{-10}\,\text{cm}^2\,\text{s}^{-1}$$

We deduce the expression for the interface concentration of carriers:

$$C^0 = \frac{1}{R_W} \times \frac{RT}{n^2 F^2} \times \frac{\delta}{D}$$

$$C^0 = \frac{1}{115} \times \frac{8.314 \times 698}{(2 \times 96\,480)^2} \times \frac{0.35 \times 10^{-4}}{8.6 \times 10^{-10}}$$

$$C^0 = 5.5 \times 10^{-5}\,\text{mol cm}^{-3}$$

4. The ASR is the total resistance of the electrode per unit surface area. The electrode resistance corresponds to the sum $A + R_2$. Because the sample is symmetric, the resistance of one electrode is half, or

$$\text{ASR} = \frac{A + R_2}{2} \times S$$

$$\text{ASR} = \frac{73.3 + 58.1}{2} \times \pi$$

$$\text{ASR} = 206.3\,\Omega\,\text{cm}^2$$

Solution 4.3 – Overpotential in an oxygen electrochemical pump

1. Defect responsible for ionic conduction in stabilized zirconia
   Schottky disorder is the dominant disorder in zirconia

$$0 \rightleftharpoons V_{Zr}^{4'} + 2V_O^{\cdot\cdot}$$

In yttria-stabilized zirconia, the reaction that introduces the dopant Y_2O_3 is

$$Y_2O_3 \longrightarrow 2Y'_{Zr} + 3O_O^X + V_O^{\cdot\cdot}$$

This doping leads to an increase in the rate of oxygen vacancies and, consequently, to a decrease in the rate of zirconium vacancies. Because of the small charge relative to the oxide ions and of the high concentration of oxygen vacancies, the electric conduction is due entirely to the oxide ions *via* a vacancy mechanism.

2. **a.** Expression for current I required to purify the gas inside the tube

Gas purification corresponds to a decrease $\Delta n_{/s}$ of the number of moles of oxygen per second of

$$\Delta n_{/s} = \frac{I}{4F} = \frac{(P_e - P_{ex})V_{/s}}{RT}$$

which gives
$$I = \frac{4F \times (P_e - P_{ex})V_{/s}}{RT}$$

b. Inserting the numerical values gives the current

$$I = \frac{4 \times 96\,480 \times (10^{-3} - 10^{-5})\,10^5 \times 5 \times 10^{-3}}{8.314 \times 913 \times 3\,600} = 7 \times 10^{-3}\,A$$

$$I = 7\,mA$$

3. The average pressure is

$$\overline{P} = \frac{10^{-3} + 10^{-5}}{2} = 5 \times 10^{-4}\,bar$$

with
$$\log \overline{P} = -3.33$$

As per figure 72, this average pressure corresponds to a normalized polarization resistance equal to $\log R_\eta = 1.85$

$$R_\eta \approx 71\,\Omega\,cm^2$$

or a polarization resistance R_p of

$$R_p = \frac{R_\eta}{S}$$

$$R_p = \frac{71}{2 \times \pi \times 20}$$

$$R_p = 0.565\,\Omega$$

The expression relating cathodic overpotential η_c to the current,

$$\eta_c = \frac{RT}{4F} \ln\left(1 - \left|\frac{I}{I_{\ell c}}\right|\right)$$

reduces to

$$\eta_c = -\frac{RT}{4F} \times \frac{I}{I_{\ell c}}$$

under the conditions $I \ll I_{\ell c}$. This allows us to deduce the polarization resistance

$$R_{\eta_c} = \frac{d\eta_c}{dI} = -\frac{RT}{4F} \times \frac{1}{I_{\ell c}}$$

$$|R_{\eta_c}| = \frac{RT}{4F|I_{\ell c}|}$$

$$|I_{\ell c}| = \frac{RT}{4FR_{\eta_c}}$$

Numerical evaluation gives

$$|I_{\ell c}| = \frac{8.314 \times 913}{4 \times 96\,480 \times 0.565} = 0.0348 \text{ A}$$

$$|I_{\ell c}| = 34.8 \text{ mA}$$

4. We can verify that, under these conditions, $|I| < |I_{\ell c}|$ or $7\text{ mA} < 34.8\text{ mA}$. If we exceed this limit, another reduction reaction is triggered. It may correspond to the reduction of the electrolyte with an increase in the electronic conductivity and the progressive blackening of the stabilized zirconia.

5. **a.** Referring to figure 72, we can write

$$\log R_\eta = -\alpha \log P_{O_2}$$

$$\alpha = -\frac{\Delta \log R_\eta}{\Delta \log P_{O_2}}$$

$$\alpha = \frac{3.8 - 0}{6 - 0.74} = 0.72$$

which gives $\qquad R_\eta = R_\eta^\circ P_{O_2}^{0.72}$

where R_η° is a constant.

Taking into account the relation obtained in question 3, we can write

$$I_{\ell c} = I_{\ell c}^\circ P_{O_2}^{-0.72}$$

where $I_{\ell c}^\circ$ is a constant.

This relation is valid for $P_{O_2} < 7.4 \times 10^{-3}$ bar.

b. In the domain of high oxygen partial pressure ($P_{O_2} > 7.4 \times 10^{-3}$ bar), for example, in air, we can equally well write

$$\log R_\eta = -\alpha' \log P_{O_2}$$

where α' is a constant. An approach identical to that following question 5(a) leads to $\qquad I_{\ell c} = I_{\ell c}^\circ P_{O_2}^{-0.46} \qquad$ where $I_{\ell c}^\circ$ is a constant.

6. a. Resistance of the electrolyte

We start by determining the value of A

$$\sigma = 1.78 \times 10^{-2} = \frac{A}{1000} e^{-\frac{0.85 \times 1.6 \times 10^{-19}}{1.38 \times 10^{-23} \times 1000}}$$

or $\qquad A = 339\,175 \ \mathrm{S\,cm^{-1}K}$

We deduce the conductivity σ at 913 K

$$\sigma = \frac{339175}{913} e^{-\frac{0.85 \times 1.6 \times 10^{-19}}{1.38 \times 10^{-23} \times 913}}$$

$$\sigma = 7.62 \times 10^{-3} \ \mathrm{S\,cm^{-1}}$$

and the resistance R of the electrolyte

$$R = \frac{1}{\sigma} \times \frac{\ell}{S}$$

$$R = \frac{1}{7.62 \times 10^{-3}} \times \frac{0.2}{2 \times \pi \times 20}$$

$$R = 0.21\,\Omega$$

b. Value of potential difference

The expression for the potential difference ΔU to apply to the terminals of the pump is

$$\Delta U = \Delta E_{th} + RI + \eta_a - \eta_c$$

where ΔE_{th} is the thermodynamic potential (at $I = 0$), η_a is the anodic overpotential, η_c is the cathodic overpotential, and I is the current in the circuit.

The thermodynamic potential ΔE_{th} is given by the expression

$$\Delta E_{th} = \frac{RT}{4F} \ln \frac{P_{int}}{P_{ext}(air)}$$

$$\Delta E_{th} = \frac{8.314 \times 913}{4 \times 96\,480} \ln \frac{5 \times 10^{-4}}{0.21}$$

$$|E_{th}| = 0.119 \ \mathrm{V}$$

and, by neglecting the anodic overpotential,

$$|\Delta U| = |\Delta E_{th}| + RI - \frac{RT}{4F}\ln\left(1 - \frac{I}{I_{\ell c}}\right)$$

$$|\Delta U| = 0.119 + (0.21 \times 7 \times 10^{-3}) - \frac{8.314 \times 913}{4 \times 96\,480}\ln\left(1 - \frac{7.05 \times 10^{-3}}{34.8 \times 10^{-3}}\right)$$

$$|\Delta U| = 0.125\,V$$

This is the value of the applied potential difference.

Solution 4.4 – Determination of exchange current

1. Figure 76 shows the electrode overpotential as a function of intensity.
   We see that the curve is linear. It can be described by the following equation:

$$\eta = R_p I \tag{1}$$

where the electrode polarization resistance R_p = 15 Ω.

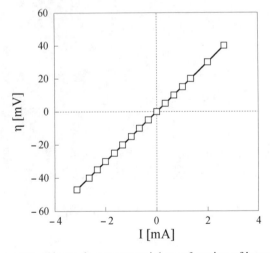

Figure 76 – Electrode overpotential as a function of intensity.

2. **a.** ▷ At the $(La_{0.8}Sr_{0.2}MnO_{3-\delta}$-YSZ) / YSZ interface, the oxygen adsorption-desorption reaction is expressed as

$$O_2 + 2s \rightleftharpoons 2O\text{-}s \tag{a}$$

where s denotes an adsorption site.

The corresponding reaction rate under the conditions given in the problem statement is

$$v_{ads} = \Gamma^2 \left[k_{ads} P_{O_2} (1 - \theta)^2 - k_{des} \theta^2 \right] \tag{2}$$

▷ At equilibrium, the speed is zero, which gives

$$\frac{1 - \theta_{eq}}{\theta_{eq}} = \left(\frac{k_{des}}{k_{ads} P_{O_2}} \right)^{1/2} \tag{3}$$

▷ The charge-transfer reaction at the triple point between gas, electrode material, and electrolyte is

$$\text{O-s} + V^{\cdot\cdot}_{\text{O,YSZ}} + 2e'_{\text{LSM}} \rightleftharpoons O^{\times}_{\text{O,YSZ}} + s \tag{b}$$

Its reaction rate v_{ct} as a function of applied potential E is

$$v_{tc} = \Gamma \left[k_{Ox} \theta C_{V^{\cdot\cdot}_O} e^{\frac{2\alpha FE}{RT}} - k_{Red}(1 - \theta) C_{O^{\times}_O} e^{\frac{2(1-\alpha)FE}{RT}} \right] \tag{4}$$

$C_{V^{\cdot\cdot}_O}$ and $C_{O^{\times}_O}$ are the respective concentrations of oxygen vacancies and oxide ions in the electrolyte.

b. We deduce the following current density passing through the electrode at potential E:

$$i = 2F\Gamma \left[k_{Ox} \theta C_{V^{\cdot\cdot}_O} e^{\frac{2\alpha FE}{RT}} - k_{Red}(1 - \theta) C_{O^{\times}_O} e^{\frac{2(1-\alpha)FE}{RT}} \right] \tag{5}$$

3. At thermodynamic equilibrium ($E = E_{th}$), the exchange-current density is i_0. The anodic current density is equal in magnitude but in the direction opposite that of the cathodic current. It may be obtained by letting the current density under polarization go to zero. We obtain

$$i_0 = 2F\Gamma k_{Ox} \theta_{eq} C_{V^{\cdot\cdot}_O} e^{\frac{2\alpha FE_{th}}{RT}} \tag{6}$$

$$i_0 = 2F\Gamma k_{Red}(1 - \theta_{eq}) C_{O^{\times}_O} e^{-\frac{2(1-\alpha)FE_{th}}{RT}} \tag{7}$$

4. a. Combining relations (5)–(7) allows us to write

$$i = i_0 \left[\frac{\theta}{\theta_{eq}} e^{\frac{2\alpha F(E - E_{th})}{RT}} - \frac{1 - \theta}{1 - \theta_{eq}} e^{-\frac{2(1-\alpha)F(E - E_{th})}{RT}} \right] \tag{8}$$

Because

$$\eta = E - E_{th} \tag{9}$$

we have

$$i = i_0 \left[\frac{\theta}{\theta_{eq}} e^{\frac{2\alpha F\eta}{RT}} - \frac{1 - \theta}{1 - \theta_{eq}} e^{-\frac{2(1-\alpha)F\eta}{RT}} \right] \tag{10}$$

b. Under conditions of negligible fractional surface coverage ($\theta \ll 1$) and with low overpotential ($\theta \approx \theta_{eq}$ and $E \approx E_{th}$), equation (10) simplifies. After expanding the exponentials, we obtain

$$i = \frac{2Fi_0}{RT}\,\eta \tag{11}$$

c. Calculating $I = i \times S$, we obtain

$$\eta = \frac{RT}{2Fi_0 S}\,I \tag{12}$$

The overpotential is a linear function of current.

5. The experimentally observed linearity of the overpotential as a function of current can be explained by the model developed in the preceding questions. This allows us to write the expressions for exchange current:

$$I_0 = i_0 \times S$$

$$I_0 = \frac{RT}{2FR_p}$$

Numerical evaluation gives

$$I_0 = \frac{8.314 \times 1020}{2 \times 96\,480 \times 15}$$

$$I_0 = 2.93\,\text{mA}$$

Solution 4.5 – Reduction of water vapor at the M / YSZ interface with M = Pt, Ni

1. Current I as a function of cathodic overpotential η (fig. 77)

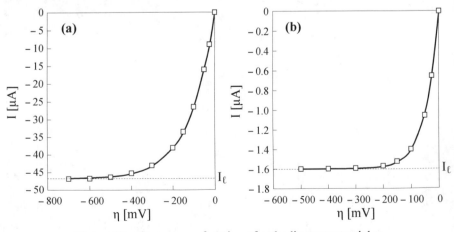

Figure 77 – Current as a function of cathodic overpotential at
(a) Pt electrode and (b) Ni electrode.

2. a. The graphical determination of the limiting currents gives

    ▷ for platinum $\qquad\qquad$ $I_\ell = -47\ \mu A$

    ▷ for nickel $\qquad\qquad$ $I_\ell = -1.6\ \mu A$

b. To obtain z, we have to derive the expression for the overpotential as a function of current.

We obtain $\qquad\qquad\qquad$ $\dfrac{\partial \eta}{\partial I} = -\dfrac{RT}{zF}\left(\dfrac{1}{I_\ell - I}\right)$

The inverse gives $\qquad\qquad$ $\dfrac{\partial I}{\partial \eta} = +\dfrac{zF}{RT}(I - I_\ell)$

The slope of the line, $\partial I/\partial \eta = f(I - I_\ell)$, which takes the value $zF/(RT)$, allow us to determine the value of z.

With a and b denoting two consecutive points of the curve $I(\eta)$, the line can be reformulated as

$$\frac{I_b - I_a}{\eta_b - \eta_a} = f\left(\frac{I_b - I_a}{2} - I_\ell\right)$$

$(I_b - I_a)/2$ is the average I_m of the current for which we calculate the slope $(I_b - I_a)/(\eta_b - \eta_a)$. The values of $\partial I/\partial \eta$ and of $I_m - I_\ell$ are listed in table 44.

Table 44 – Detail of the calculation of the function $\partial I/\partial \eta = f(I_m - I_\ell)$.

For the Pt / YSZ interface									
$10^6\ \partial I/\partial\eta\ [A\,V^{-1}]$ 360	280	210	140	90	50	22	11	4	1
$10^6\ (I_m - I_\ell)\ [A]$ 42.5	34.5	25.75	17	11.25	6.5	2.9	1.25	0.5	0.25
For the Ni / YSZ interface									
$10^6\ \partial I/\partial\eta\ [A\,V^{-1}]$ 26.04	16.12	6.86	2.48	0.94	0.23	0.05	0.02		
$10^6\ (I_m - I_\ell)\ [A]$ 1.274	0.747	0.374	0.141	0.055	0.02	0.006	0.003		

The functions are shown in figure 78 for (a) the Pt / YSZ interface and (b) the Ni / YSZ interface.

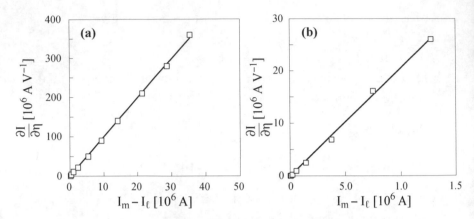

Figure 78 – Graph of $\partial I/\partial\eta = f(I_m - I_\ell)$ for
(a) Pt/YSZ interface and (b) Ni/YSZ interface.

The slopes of the lines are obtained by linear regression. For platinum, we find

▷ $\dfrac{zF}{RT} = 8.28$

   from which $z = \dfrac{8.314 \times (860 + 273) \times 8.28}{96\,480}$

$$z = 0.81$$

▷ and for nickel, we find $\dfrac{zF}{RT} = 20.54$

   from which $z = \dfrac{8.314 \times (860 + 273) \times 20.54}{96\,480}$

$$z = 2$$

Taking $z = 1$ as a first approximation, we can propose the following mechanism for platinum:

$$H_2O_{(g)} + O^{2-}_{surface} \longrightarrow 2OH^-_{(ads)}$$

$$V_O^{\cdot\cdot} + OH^-_{(ads)} + e \longrightarrow O_O^\times + \tfrac{1}{2}H_{2(g)}$$

For nickel, where $z = 2$, we can propose

$$H_2O_{(g)} \longrightarrow H_2O_{(ads)}$$

$$V_O^{\cdot\cdot} + H_2O_{(ads)} + 2e \longrightarrow O_O^\times + H_{2(g)}$$

Solution 4.6 – Hydrogen oxidation at Ni / YSZ interface

1. To calculate the oxygen partial pressure, we begin by writing the equilibrium reaction for forming water vapor:

$$H_{2(g)} + \tfrac{1}{2}O_{2(g)} \rightleftharpoons H_2O_{(g)}$$

whose change in standard free enthalpy $\Delta_r G_T^\circ$ and in the equilibrium constant K_{eq} are respectively given by

$$\Delta_r G_T^\circ = \Delta_r H_T^\circ - T\Delta_r S_T^\circ \quad \text{and} \quad K_{eq} = \frac{P_{H_2O}}{P_{H_2} P_{O_2}^{\frac{1}{2}}}$$

where the partial pressures P_i are expressed in bars.

We obtain

$$P_{O_2} = \left(\frac{P_{H_2O}}{P_{H_2}}\right)^2 e^{-\frac{2(\Delta_r H_T^\circ - T\Delta_r S_T^\circ)}{RT}}$$

Numerical evaluation gives

$$P_{O_2} = \left(\frac{1}{15.45}\right)^2 e^{-\frac{2(224\,205 + 1\,248 \times 45.73)}{8.314 \times 1\,248}}$$

$$P_{O_2} = 4.3 \times 10^{-17} \text{ bar}$$

2. **a.** The anode behavior can be described by the following reactions:

$$H_{2(g)} \rightleftharpoons H_{2(ads)}$$

$$H_{2(ads)} + O_O^\times \longrightarrow H_2O_{(ads)} + 2e + V_O^{\cdot\cdot}$$

$$H_2O_{(ads)} \rightleftharpoons H_2O_{(g)}$$

b. ▷ If the oxidation kinetics is limited by the charge-transfer step, the current-overpotential characteristic can be described by the Butler-Volmer equation

$$i = i_0 \left(e^{\frac{\alpha nF}{RT}\eta} - e^{-\frac{(1-\alpha)nF}{RT}\eta}\right)$$

A least squares fit gives the following values for i_0 and α:

$$i_0 = 8.1 \text{ Acm}^{-2}$$

and $$\alpha = 0.7$$

▷ The experimental variation of the current density as a function of overpotential for this electrode is expressed as

$$i = 8.1\left(e^{\frac{0.7 \times 2 \times 96\,480}{8.314 \times 1248}\eta} - e^{-\frac{(1-0.7) \times 2 \times 96\,480}{8.314 \times 1248}\eta}\right)$$

or
$$i = 8.1\left(e^{13.02\eta} - e^{-5.58\eta}\right)$$

c. The experimental results and the theoretical curve for the anodic over-potential as a function of the logarithm of current density are shown in figure 79.

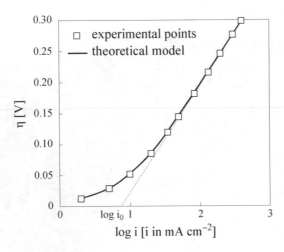

Figure 79 – Anodic over-potential as a function of the logarithm of current density.

The agreement between theory and experiment confirms the hypothesis that the oxidation kinetics for hydrogen in this electrode and in these physico-chemical conditions is limited by the charge-transfer step.

d. At high current densities, the anodic overpotential as a function of the logarithm of current density is linear. The electrode thus follows a Tafel relationship.

$$\log i = \log i_0 + \frac{\alpha n F}{2.3 RT}\eta$$

$$\log i = \log 8.1 + \frac{0.7 \times 2 \times 96\,480}{2.3 \times 8.314 \times 1248}\eta$$

$$\log i = 0.91 + 5.66\,\eta$$

Chapter 5

Applications

Course notes

5.1 – Electrochemical sensors

5.1.1 – Definition and characteristics

An electrochemical sensor is a device that is capable of detecting variations in quantities related to a chemical species in a given phase (e.g., concentration, partial pressure, molar fraction). Such sensors generally consist of an active layer that can recognize and interact with the species and that is associated with a transducer capable of transforming the interaction into an electrical signal. The signal may be potentiometric, amperometric, conductometric, … The quantities measured are the concentrations (activities) of ionic species (H_3O^+, Na^+, Cl^-) and the partial pressures of gaseous species (O_2, H_2, CO_2, H_2O vapor, …).

Electrochemical sensors fulfill sufficiently well the requisite conditions in terms of stability, selectivity, response time, and reversibility. On the practical and economic level, they are robust, simple, inexpensive, convenient, and have an acceptable lifetime. They are used in particular for

▷ industrial-process engineering: metallurgy, plastics processing;

▷ the environment: analysis of effluents, water, soil, and air;

▷ automobile industry: battery charge level, analyze exhaust gas;

▷ medicine: diagnostics;

▷ security: leaks of toxic or noxious gases.

© Springer Nature Switzerland AG 2020
A. Hammou and S. Georges, *Solid-State Electrochemistry*,
https://doi.org/10.1007/978-3-030-39659-6_5

In what follows, we restrict ourselves to developing the principles that govern the operation of potentiometric, amperometric, conductimetric and coulometric gas sensors.

5.1.2 – Potentiometric sensor for gas analysis

The principle is based on exploiting the equilibria at the gas/electrode/solid electrolyte interfaces. The sensor invokes an electrochemical chain with two electrodes separated by a solid electrolyte. The first electrode is in contact with the gaseous species to be analyzed and the second serves as a reference electrode. The sensor works by measuring the apparent open-circuit potential difference ΔE between the chain terminals. ΔE is a function of temperature and of the partial pressure of the gaseous species to be analyzed within the mix of gases. An oxygen (O_2) sensor is taken as an example in what follows. The electrochemical chain considered is

$$P_{O_2}^{(1)}, Me \,/\, \text{solid electrolyte (SE)} \,/\, Me, P_{O_2}^{(2)}$$
$$(1) \hspace{5cm} (2)$$

where Me is an inert metal and the solid electrolyte (SE) is an oxide ion conductor. Electrode (1) is the measurement electrode (Mes) and electrode (2) is the reference electrode (Ref).

The same equilibrium reaction occurs at electrodes (1) and (2):

$$\tfrac{1}{2}O_2 + 2e_{Me} \rightleftharpoons O_{SE}^{2-}$$

The chemical potential for oxygen

$\triangleright$ at electrode (1) is

$$\frac{\mu_{O_2}^{(1)}}{2} = \frac{\mu_{O_2}^{\circ}}{2} + \frac{RT}{2}\ln P_{O_2}^{(1)} = \tilde{\mu}_{O^{2-}}^{(1)} - 2\mu_e^{(1)} + 2F\varphi^{(1)}$$

$\triangleright$ at electrode (2) is

$$\frac{\mu_{O_2}^{(2)}}{2} = \frac{\mu_{O_2}^{\circ}}{2} + \frac{RT}{2}\ln P_{O_2}^{(2)} = \tilde{\mu}_{O^{2-}}^{(2)} - 2\mu_e^{(2)} + 2F\varphi^{(2)}$$

with the equilibrium $\tilde{\mu}_{O^{2-}}^{(1)} = \tilde{\mu}_{O^{2-}}^{(2)}$ and, for a given metal electrode, $\mu_e^{(1)} = \mu_e^{(2)}$.
We thus deduce the potential difference ΔE between the sensor terminals:

$$\Delta E = \varphi^{(2)} - \varphi^{(1)} = -\frac{RT}{4F}\ln \frac{P_{O_2}^{(1)}}{P_{O_2}^{(2)}}$$

$$\Delta E = -\frac{RT}{4F} \ln \frac{P_{O_2}^{(Mes)}}{P_{O_2}^{(Ref)}}$$

5.1.3 – Amperometric sensor

The working principle of this sensor is based on exploiting the diffusion-limited current of the electroactive species (Ox or Red) at an electrode/solid electrolyte interface. This current is proportional to the concentration or the molar fraction of the species being analyzed (see Chapter 4, Thermodynamics and electrochemical kinetics). In practice, a diffusion barrier (small holes or porous layer) is created in the device, which limits the diffusion of active species. The sensor is often calibrated beforehand.

In what follows, we use the example of a gas sensor designed to detect the concentration of species X_2 as it undergoes the reduction reaction

$$\tfrac{1}{2}X_2 + e \rightleftharpoons X_{SE}^- \qquad \text{SE: solid electrolyte}$$

The sensor response is shown in figure 80.

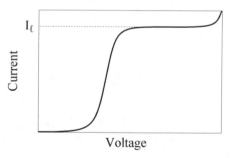

Figure 80 – General form of X_2-reduction current as a function of electrode voltage for the case of a sensor based on a diffusion barrier.

I_ℓ is the diffusion-limited current of X_2. It is a function of the molar fraction x_{X_2} of X_2 and is written

▷ for normal diffusion: $I_\ell = K \ln(1 - x_{X_2})$
▷ for Knudsen diffusion: $I_\ell = K'P_t \ln x_{X_2}$
   where P_t denotes the total pressure of the gas and K and K' are constants.

Note – If the gaseous mix contains several species that can be reduced at the cathode, the sensor theoretically reveals the same number of limiting currents. Each current is associated with a given species.

The sensor is developed in particular for analyzing exhaust gas.

5.1.4 – Coulometric sensor

The working principle of a coulometric sensor is based on the complete consumption by electrolysis of the species being analyzed. The amount of substance due to this species is directly related to the amount of charge $Q = \int I dt$ that passes through the electrode. This type of sensor requires calibration beforehand and cannot make continuous measurements.

5.1.5 – Conductometric sensor for gas analysis

The conductometric sensor for gas analysis normally relies upon an ionic non-stoichiometric solid. Its working principle is based on the electronic conductivity of the solid as a function of the composition of the surrounding gas. For example, to measure the amount of oxygen in a mix of gases, we can use SnO_2, $BaTiO_3$, or solid solutions based on TiO_2, such as TiO_2-CoO-BaO.

5.2 – Electrochemical generators

5.2.1 – Definition and characteristics

◇ *Electrochemical generator: Definition, capacity, and theoretical energy*

The electrochemical generator enables the direct transformation of chemical energy into electrical energy. Schematically, it consists of an electrochemical chain that has two electrodes with differing electric potentials and that are separated by an electrolyte. The electrode with the higher potential, denoted $E_1 = E_+$, is called positive and is the site of the equilibrium

$$Ox_1 + ne \rightleftharpoons Red_1$$

The other electrode is called negative, has a potential $E_2 = E_-$, and is the site of the equilibrium $\qquad Ox_2 + ne \rightleftharpoons Red_2$

The result is an electromotive force ΔE (or voltage U) that is always positive and is given by $\qquad U = \Delta E = E_+ - E_-$

U is related to the free enthalpy of the overall working reaction of generator

$$Ox_1 + Red_2 \longrightarrow Red_1 + Ox_2$$

by the expression $\qquad\qquad U = -\dfrac{\Delta_r G}{nF}$

◇ *Theoretical capacity and nominal capacity*

Several definitions exist for theoretical capacity. In terms of conceiving a generator, we define the specific capacity C_m and the volumetric capacity C_V with respect to a single electrode [(1) or (2)] as follows:

$$C_m = \frac{nF}{M} \quad \text{and} \quad C_V = \frac{nF}{V_m}$$

n is the number of electrons involved in the electrode reaction and M and V_m are the molar mass and molar volume, respectively, of the reactant (Ox for the cathode and Red for the anode). A second definition takes into account the complete chain. In this case, we have

$$C_m = \frac{nF}{M_{Ox_1} + M_{Red_2}} \quad \text{and} \quad C_V = \frac{nF}{V_{m,Ox_1} + V_{m,Red_2}}$$

The specific and volumetric capacities are normally expressed in $mA\,h\,g^{-1}$ and $mA\,h\,cm^{-3}$, respectively.

The nominal capacity C_{nom} is the capacity indicated on the generator by the manufacturer. It is always less than the theoretical capacity.

◇ *Theoretical energy of a generator*

The theoretical energy is the product of the theoretical capacity and the theoretical emf
$$E_{th} = C_{th} \times U_{th}$$

For the capacity, we define a specific theoretical energy and a volumetric theoretical energy.

5.2.2 – Discharge and (re)charge of electrochemical generators

In the discharge regime, a generator delivers a dc current I as output. This current comes from the spontaneous reduction and oxidation reactions at the positive and negative electrodes, respectively. Taking as an example the reactions discussed above, the potential difference U_{av} "available" to the user is

$$U_{av} = \Delta E - RI - \eta_a^{(-)} - |\eta_c^{(+)}|$$

where R is the sum of the electrolyte resistance and the contact resistance and $\eta_a^{(-)}$ and $\eta_c^{(+)}$ are the overpotentials of the anode and cathode, respectively.

The manufacturer often sets a maximum potential $U_{av,max}$ for the generator.

◇ *Coulombic efficiency*

The coulombic efficiency is the ratio $\rho = \frac{C_{disc}}{C_{nom}}$ where C_{disc} denotes the capacity recuperated at the maximum potential.

Charging or recharging a generator consists of supplying it with electrical energy so that the reactions at the electrodes are the opposite of those that occur during discharge. The generator in this case behaves as a receiver and the minimal potential U_c to apply to its terminals is

$$U_c = \Delta E + RI + \eta_a^{(+)} + |\eta_c^{(-)}|$$

NB – This equation is the same as that for the potential to apply to the terminals of an electrolyzer.

◇ *Current-voltage and power-current characteristics*

Figure 81 shows the general form of the curves representing the potential U and the power P delivered to the terminals of a generator as a function of current I. Note in particular the maximum power P_{max} at the current $I_{P_{max}}$. In general, this point is chosen by the user to obtain the optimal operating characteristics of the generator.

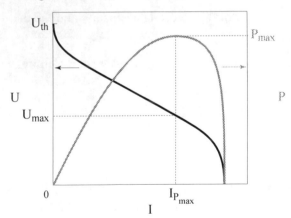

Figure 81 – Typical voltage-current and power-current curves for an electrochemical generator.

5.2.3 – Primary batteries, fuel cells, and secondary batteries

Electrochemical generators are generally grouped into two main classes: primary batteries (non-rechargeable) and fuel cells on the one hand and secondary batteries (rechargeable) on the other hand. Only discharge is possible for primary batteries.

◇ Primary and secondary batteries

The most studied systems are primary and secondary batteries that combine lithium in metallic form or in solid solution (lithium-ion) as negative electrode and an insertion material as positive electrode; the two are separated by a lithium-ion solid electrolyte conductor. The objective is to create generators that deliver voltages on the order of 3 V with specific capacities around 300 mAh g^{-1}. We limit ourselves to the two following examples:

▷ Li/V_2O_5, with $0 < x < 1$

$$xLi + V_2O_5 \rightleftharpoons Li_xV_2O_5$$

▷ $LiMVO_4/LiMn_{2-y}M_yO_4$, with M = Ni, Co, Cd, Zn

$$LiMVO_4 + LiMn_{2-y}M_yO_4 \rightleftharpoons Li_{1-x}MVO_4 + Li_{1+x}Mn_{2-y}M_yO_4$$

A promising positive electrode is $LiFePO_4$.

◇ Fuel cells

Unlike the majority of the classic primary and secondary batteries, fuel cells are "open" systems where the electrodes are continuously supplied with an active substance while in operation. We can cite as an example polymer-electrolyte fuel cells, such as *Nafion*-type fuel cells, which are proton-exchange membrane fuel cells (PEMFCs), and solid oxide fuel cells (SOFCs), such as those that use stabilized zirconia. The fuel in question is hydrogen and is generally obtained by external or internal reforming of methane, and the oxidizer is oxygen from the ambient air. The overall reaction in a fuel cell is the formation reaction of water

$$H_2 + \tfrac{1}{2}O_2 \longrightarrow H_2O$$

The elementary cell may be considered as an oxygen concentration cell whose emf is given by the Nernst equation.

PEMFCs operate around 80 °C and presently use *Nafion* as electrolyte. The electrodes are made of carbon that encloses Pt nanoparticles, which act as electrocatalysts for the cathodic and anodic reactions.

SOFCs operate between 600 and 800 °C. The electrolyte consists of yttria-stabilized zirconia, with the formula $(ZrO_2)_{0.92}(Y_2O_3)_{0.08}$. The cathode material is a solid solution of strontium-doped lanthanum manganite with the formula $La_{1-x}Sr_xMnO_{3-\delta}$. The anode is made of an yttria stabilized nickel-zirconia cermet containing 40% nickel by volume. It operates at high temperature, which is exploited for cogeneration; in other words, to simultaneously produce electricity and heat.

Exercises

Exercise 5.1 – Determination of standard free enthalpy of formation for AgCl

To determine the standard free enthalpy $\Delta_f G°$ of formation of solid silver chloride AgCl, we use the following electrochemical chain:

$$C_1 / Ag / AgCl / C_2, Cl_2$$
$$\quad \alpha \quad \beta \quad \gamma$$

where C_1 and C_2 represent the graphite electrodes, which are purely electronic conductors.

1. Use the electrochemical potential of the species to write the expression relating the standard free enthalpy $\Delta_f G°$ for the formation reaction of AgCl, with a potential difference ΔE measured between the terminals of the electrochemical chain. Assume that all thermodynamic equilibria hold.

2. A potential difference of 924 mV is measured between the terminals of the electrochemical chain at 350 °C and under a chlorine partial pressure of 5×10^4 Pa. Deduce the standard free enthalpy of formation of AgCl in these conditions.

3. Compare the result with that calculated by using the thermodynamic data.

Data

	$Ag_{(s)}$	$Cl_{2(g)}$	$AgCl_{(s)}$
Standard enthalpy of formation $\Delta_f H°$ [kJ mol^{-1}]	–	–	– 127.1
Absolute standard entropy $S°$ [J mol^{-1} K^{-1}]	42.5	223	96.2

Exercise 5.2 – Measurement of thermodynamic quantities of metal fluorides

We want to prepare a solid electrolyte based on calcium fluoride CaF_2. The dominant disorder in this ionic compound is anionic Frenkel disorder.

1. Give the reaction that involves anionic Frenkel disorder.

2. To increase the ionic conductivity of CaF_2, we dope it by adding sodium fluoride NaF or yttrium fluoride YF_3. Given that the cationic species relative to the dopant occupies a substitution site, write for each case the reaction for introducing the dopant into CaF_2.

3. CaF_2 reacts with water vapor to liberate gaseous hydrogen fluoride HF. State the reaction given that the atomic species are introduced into substitution sites.

4. Give two examples of complexes that may be formed in each type of solid solution obtained in questions 2 and 3. Give the corresponding equilibrium reactions and the effective charge of the complex obtained.

5. The solid solution $(CaF_2)_{0.99}$ $(NaF)_{0.01}$ is a solid that conducts *via* fluoride ions F^-. Use this to determine the free enthalpy of formation of a metal fluoride with formula MF_2 at 1 000 K. To do this, consider the following electrochemical chain:

$$Pt, Ni\text{-}NiF_2 \, / \, (CaF_2)_{0.99}(NaF)_{0.01} \, / \, M\text{-}MF_2, Pt$$
$$\qquad I \qquad\qquad\qquad\qquad\qquad\qquad II$$

The fluorine pressure created by the $Ni\text{-}NiF_2$ mixture in compartment I is 7.4×10^{-39} bar at 1 000 K. The electromotive force ΔE measured between the terminals of the chain is 0.794 V, with the nickel electrode being positive.

a. Calculate the fluorine partial pressure imposed by the $M\text{-}MF_2$ pair.
b. Deduce the standard free enthalpy of formation of MF_2 at 1 000 K.
c. With the help of table 45, identify the metal M.

Table 45 – Standard free enthalpies of formation for various metal fluorides.

Compound	CuF_2	MnF_2	FeF_2	MgF_2	CdF_2	NiF_2
$\Delta_f G°$ [kJ mol^{-1}]	– 627	– 883	– 789	– 1 002	– 793	– 734

Exercise 5.3 – Measurement of O^{2-} ion activity in a molten salt

We wish to measure the activity of O^{2-} ions in a molten salt. A schematic of the measurement cell appears in figure 82.

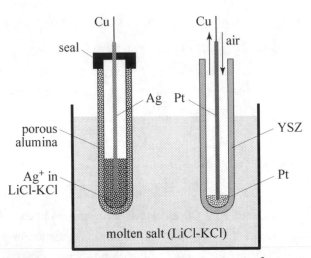

Figure 82 – Schematic of cell used for measuring O^{2-} activity
in molten salt LiCl-KCl (from Tremillon & Picard, 1983).

This cell uses the following electrochemical chain:

$$\begin{array}{cccccccc}
\alpha & \beta & & \gamma & \delta & & \varepsilon & \lambda & \chi \\
\end{array}$$

$$\text{Cu} / \text{Ag} / \text{Ag}^+ \text{ in LiCl-KCl} / \text{porous alumina} / O^{2-} \text{ in LiCl-KCl} / \text{YSZ} / \text{Pt,O}_2 / \text{Cu}$$

$$\begin{array}{cccccccc}
(1)(2) & & (3) & & (4) & & (5) & (6) \ \ (7) \ \ (8)
\end{array}$$

where α, β, γ, ..., χ represent the interfaces and (1), (2), (3), ..., (8) represent
the phases.

The cell consists of

   ▷ a reference electrode made of the first species (Ag$^+$/Ag) (interface β) obtained by dissolving a certain amount of AgCl in the molten salt LiCl-KCl,
denoted "int. ref.,"

   ▷ a gaseous measuring electrode (interface λ) obtained by depositing a platinum coating (7) on the inside of a stabilized zirconia tube (YSZ), which is
in contact with the ambient air,

   ▷ the stabilized zirconia membrane (6), which is sensitive to O^{2-} ions (interface ε),

   ▷ a liquid junction within a porous alumina tube (4) between the reference
salt (3) and the salt being studied (5).

The emf of the chain is $\qquad \Delta E = \varphi^8 - \varphi^1$

where φ is the electric potential.

The following hypotheses allow the calculation of the open-circuit emf of the chain:

▷ an equilibrium exists at each interface, taking into account the dominant carriers in each solid phase,

▷ the electric potential of each species is constant within each solid phase that conducts *via* a single species (electrons or ions),

▷ the stabilized zirconia (ZrO_2–Y_2O_3 8 mol%, YSZ) is considered a strictly ionic conductor,

▷ the ionic junction between phases (3) and (5) is implemented with porous alumina (4). The molten salts on each side of the alumina membrane are very similar compounds with a high concentration of ions. We can thus consider that the potential difference ($\varphi^5 - \varphi^3$) across the junction is negligible.

1. Represent schematically the electric potential within the electrochemical chain.

2. Determine the emf of the sensor as a function of the concentration of O^{2-} ions dissolved in the molten salt and, more precisely, of pO^{2-} (*i.e.*, $-\log[O^{2-}]$).

3. We add zinc oxide to the molten salt and do the same experiment with lithium oxide Li_2O. The emf of the sensor is given for each case in figure 83 as a function of pO^{2-}.
   a. Explain the origin of the two straight lines.
   b. Considering only the case of ZnO, compare the slope of the line obtained in question 2.
   c. Give the solubilization reactions of the oxides, the corresponding equilibrium constants, and the expressions relating the solubilities to the constants.
   d. Evaluate the solubilities of ZnO and Li_2O.

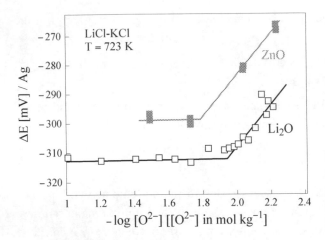

Figure 83 – emf as a function of the concentration of added oxide ions (Li$_2$O, ZnO) (from Schenin-King, 1994).

Exercise 5.4 – Calculation of equilibrium constants for defect formation in Cu$_2$O

Copper oxide Cu$_2$O is a known p-type semiconductor and has copper vacancies and holes as its main defects. Analyses of the departure from stoichiometry at the copper site and measurements of the diffusion tracer ^{18}O have revealed the presence of neutral copper vacancies V_{Cu}^{x} and singly ionized copper vacancies V_{Cu}' on the one hand, and singly ionized oxygen interstitials O_i' on the other hand. Table 46 gives the electrical conductivity of Cu$_2$O as a function of the oxygen partial pressure at 1 000 and 1 100 °C.

Table 46 – Electrical conductivity of Cu$_2$O as a function of oxygen partial pressure at 1 000 and at 1 100 °C (from Peterson & Wiley, 1984).

log P$_{O_2}$ [bar]	$\sigma_{1000\,°C}$ [S cm^{-1}]	$\sigma_{1100\,°C}$ [S cm^{-1}]
– 1	7.108	12.420
– 1.5	5.754	10.000
– 2	4.677	8.273
– 2.5	3.890	6.844
– 3	3.236	5.662
– 3.5	2.661	4.684
– 4	2.213	3.981
– 4.5	1.884	3.384
– 5	1.641	–

1. Propose a model that explains all the defects supported by experimental evidence. To do this, write the reactions involved and the corresponding equilibrium constants.

2. Establish a relationship between the electrical conductivity, the constants of formation of mobile structure defects, the hole mobility, and the oxygen partial pressure.

3. **a.** By using least squares fitting and starting from the relationship obtained in question 2, determine the expressions for the conductivity of Cu_2O as a function of oxygen partial pressure at 1 000 and 1 100 °C.

   b. On a logarithmic scale, represent the experimental and theoretical conductivity of Cu_2O as a function of oxygen partial pressure at 1 000 and 1 100 °C.

4. Based on the equations obtained for question 3(a) and considering a constant hole mobility of $1 \ m^2 V^{-1} s^{-1}$, calculate the values of
   a. the equilibrium constant for the formation reaction for singly ionized copper vacancies (V'_{Cu}) at 1 000 and 1 100 °C,
   b. the equilibrium constant for the formation reaction for singly ionized oxygen interstitials (O'_i) at 1 000 and 1 100 °C.

Exercise 5.5 – TiS$_2$: insertion material

1. Define and explain the role of an insertion material (IM) in electrochemistry.

2. Give several examples, other than TiS_2, of insertion materials for lithium that are used for electrodes.

3. List three conditions necessary for inserting lithium into an insertion material.

4. Consider the following lithium battery:

$$Li \ / \ ionic \ Li^+ \ conducting \ electrolyte \ / \ TiS_2$$

   a. Write the electrochemical reactions at the electrodes and the operating reaction of the generator corresponding to a discharge rate x, and specify the formal charges of the cations at the cathode.

   We consider in what follows that insertion in TiS_2 is limited only by ions. The insertion rate of lithium into TiS_2 as a function of the potential at 298 K is given in table 47.

Table 47 – Insertion rate of lithium
into TiS$_2$ as a function of potential at 298 K.

x	0.01	0.1	0.2	0.3	0.4	0.5	0.6	0.7	0.8	0.9	0.99
emf [V]	2.49	2.4	2.36	2.31	2.26	2.215	2.18	2.13	2.07	1.99	1.86

By using the thermodynamic model

$$E_{Li} = E° - \frac{RT}{F} \ln \frac{x}{x_m - x}$$

where x is the charge rate and x_m is the maximum charge rate ($x_m = 1$),

▷ give the conditions for which this equation may be derived,

▷ express the potential of a TiS$_2$ electrode as a function of the discharge
rate x,

▷ based on the results given in table 47 and on the thermodynamic
model, plot the experimental curve for the open-circuit emf as a function of the insertion rate x. Discuss any similarities between the two.

c. To better understand the experimental data, we add to the model a thermodynamic energy term E_{int} for the interaction between lithium ions.
The model then takes the form

$$E_{Li} = E° - \frac{RT}{F} \ln \frac{x}{x_m - x} - z \frac{E_{int}}{F} \frac{x}{x_m}$$

where z is the number of nearest neighbors of Li$^+$ ($z = 6$).

▷ Determine the new equation.

▷ Determine the value for E_{int} that leads to the best agreement with experimental data and represent the situation graphically. Discuss any
agreement between the experimental points and the theory.

5. TiS$_2$ crystallizes in a hexagonal structure. Figure 84 shows TiS$_2$ as seen
looking down the c axis. Show the position of the inserted Li$^+$ ions. Calculate
the variation ΔV [%] of the volume of TiS$_2$ that corresponds to a unity discharge rate.

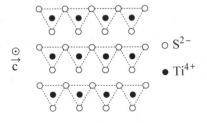

o S^{2-}

• Ti^{4+}

Figure 84 – Representation
of [001] projection of TiS$_2$.

Data

Material	TiS$_2$	LiTiS$_2$
a [Å]	3.407	3.455
c [Å]	5.697	6.195

Exercise 5.6 – Chlorine sensor based on doped strontium chloride

A – Study of strontium chloride SrCl$_2$

Figure 85 shows the conductivity in Arrhenius coordinates of pure SrCl$_2$ and of SrCl$_2$ doped by KCl.

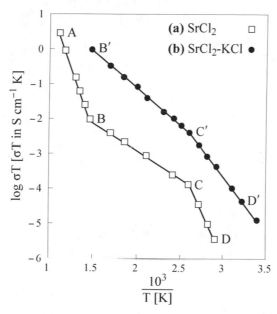

Figure 85 – Conductivity in Arrhenius coordinates of
(a) pure SrCl$_2$ and (b) SrCl$_2$ doped with potassium chloride KCl.

1. Given that the dominant disorder is anionic Frenkel disorder,
   a. Write the equilibrium for structure defects involved and the associated equilibrium constant K_{AF}. The expression for this constant as a function of temperature is

$$K_{AF} = 9 \times 10^{-4} e^{-\frac{1.6\,eV}{kT}}$$

   b. Identify the various temperature domains that appear in figure 85(a).

c. The strontium is supposedly immobile. The electrochemical mobilities of the mobile species $V_{Cl}^{\cdot}$ and Cl_i' are given by

$$\tilde{u}_{V_{Cl}^{\cdot}} = \frac{6 \times 10^{-3}}{T} e^{-\frac{0.4\,eV}{kT}}$$

$$\tilde{u}_{Cl_i'} = \frac{36}{T} e^{-\frac{1\,eV}{kT}}$$

At what temperature do the two species have the same mobility?

d. Given that the conductivity of particle i is given by

$$\sigma_i = z_i^2 F^2 \tilde{u}_i[i]$$

▷ express the total conductivity of $SrCl_2$ for the conditions given in question 1(c) and assuming that it is purely ionic.

▷ Calculate the total conductivity.

2. a. Write the dissolution reaction for KCl in $SrCl_2$ given that the potassium substitutes for strontium.

b. Identify the various temperature domains that appear in figure 85(b).

3. a. Make qualitative Brouwer diagrams of pure $SrCl_2$ and of the solid solution $SrCl_2$-KCl with "a" denoting the doping level.

b. Justify the use of the solid solution $SrCl_2$-KCl as electrolyte for making a chloride sensor.

4. We introduce silver chloride AgCl into the solid solution $SrCl_2$-KCl.

a. Write the reaction for doping the solid solution $SrCl_2$-KCl with silver chloride AgCl. We assume that silver occupies interstitial positions and that chlorine occupies normal positions.

b. Denote by b the concentration of AgCl in the resulting solid solution. Express the total ionic conductivity σ_t' for the case where $b = a = \sqrt{K_{AF}}$ and taking into account the identical mobilities of $V_{Cl}^{\cdot}$ and Cl_i'.

c. Calculate σ_t' for these conditions.

Data

$$\left| \; \tilde{u}_{Ag^+} = \frac{6 \times 10^2}{T} e^{-\frac{1\,eV}{kT}} \right.$$

B – Study of chlorine sensor

We consider the following isothermal electrochemical chain:

$$Cl_2(P_1), Me \; / \; SrCl_2\text{-}KCl \; / \; Me, Cl_2(P_2)$$
$$(1) \qquad\qquad\qquad\qquad (2)$$

where P_1 and P_2 denote the partial pressure of chlorine at electrodes (1) and (2), respectively.

1. Write the electrochemical reaction at the electrodes.

2. With the help of the electrochemical potentials of the species involved, derive the expression relating the emf at the chain terminals to the chlorine partial pressure and to the temperature.

3. We build the sensor shown schematically in figure 86 where ruthenium dioxide RuO_2 is an electronically conducting catalyst. Ag and AgCl are distinct solid phases.

   a. Derive the expression relating the emf at the chain terminals to the chlorine partial pressure.

   b. Specify the role of the Ag-AgCl mixture.

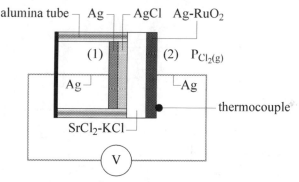

Figure 86 – Schematic representation
of a chlorine sensor (from Déportes *et al.*, 1994).

4. Table 48 gives the sensor emf as a function of temperature for a Cl_2 partial pressure of 1 bar.

Table 48 – emf of chlorine sensor as a function
of temperature for a Cl_2 partial pressure of 1 bar.

T [°C]	115.5	165.7	203.2	262.4	311.7	376.7	434.9
emf [mV]	1 100.0	1 070.4	1 044.0	1 010.2	978.5	941.5	909.9

   a. Plot the emf as a function of absolute temperature T and deduce the relationship $\Delta E = f(T)$.

b. Verify that the data in the table are consistent with the theoretical expression for emf.

c. Deduce the standard free enthalpy of formation of solid AgCl at 350 °C.

Exercise 5.7 – CO₂ sensor (a)

Figure 87 shows a schematic representation of a CO_2 sensor including a Na^+ conducting ceramic.

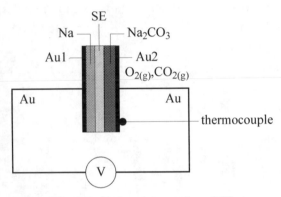

Figure 87 – Schematic representation of CO_2 sensor.

SE is a Na^+ conducting solid electrolyte and (α), (β), and (γ) represent respectively the interfaces Au1/Na, Na/SE, and SE/Na₂CO₃. The potential is measured between the gold collectors. The sodium layer is 1 μm thick with a surface area of 0.4 cm² and serves as reference. The Na₂CO₃ layer is porous to allow gas exchange with the measurement electrode.

1. a. Write down the electrochemical chain used for this sensor.

b. Give the analytic expression for the emf ΔE delivered by the sensor as a function of standard free enthalpy $\Delta_r G_T^\circ$ of the following reaction:

$$2Na + \tfrac{1}{2}O_2 + CO_2 \rightleftharpoons Na_2CO_{3(s)}$$

We associate the activity of the gaseous constituents to their partial pressure in bars.

2. Calculate the emf for a carbon dioxide partial pressure $P_{CO_2} = 10^{-3}$ bar.

Data

$\Delta_r G_T^\circ = -742.37 + 0.276\,T$ in kJ mol⁻¹ and $P_{O_2} = 0.21$ bar

3. Taking the sensor resistance to be negligible, evaluate the lifetime of the reference (Na) if, to measure the emf, we use a voltmeter with a small input impedance ($10^7 \, \Omega$) for $P_{CO_2} = 10^{-5}$ bar.

Data

> Sodium molar mass $M_{Na} = 23$ g mol^{-1}
> Sodium density $\rho_{Na} = 0.97$ g cm^{-3}

4. We define the isoelectric point of the sensor as the pressure P_{CO_2} that renders the emf independent of measurement temperature. What is this pressure? Does it fall within the range of pressure of CO_2 in the ambient air?

5. We replace the sodium layer, which is difficult to fabricate, by a reference system that involves the solid solution $(AgBr)_{0.99}(NaBr)_{0.01}$. Under the measurement conditions, this solution is an ionic conductor.

   a. Write down the electrochemical chain.
   b. Give the analytic expression for the emf at the sensor terminals.
   c. Discuss the evolution of the emf when we use a voltmeter with a small input impedance and qualitatively compare the result with that of question 3.

Exercise 5.8 – CO$_2$ sensor (b)

We consider a carbon dioxide CO_2 potentiometric sensor based on the following electrochemical chain:

$$CO_2,O_2 \, / \, LSM,Na_2Ti_6O_{13}\text{-}TiO_2 \, / \, NASICON \, / \, Na_2CO_3,Au \, / \, O_2,CO_2$$

$$\underbrace{\phantom{CO_2,O_2 \, / \, LSM,Na_2Ti_6O_{13}\text{-}TiO_2}}_{\substack{\text{reference electrode} \\ \text{(RE)}}} \quad \underbrace{}_{\text{electrolyte}} \quad \underbrace{}_{\substack{\text{measurement electrode} \\ \text{(ME)}}}$$

where LSM is an oxide-based electrode material with dominant electronic conduction and the formula $La_{0.8}Sr_{0.2}MnO_{3-\delta}$ and $Na_2Ti_6O_{13}\text{-}TiO_2$ is a mix of two solid phases. NASICON is a Na^+ ionic conducting solid solution with the formula $Na_3Zr_2Si_2PO_{12}$, and Na_2CO_3 is the sodium carbonate in the measurement electrode.

We assume that the solid phases are pure.

1. Write the equilibria that occur at the reference and measurement electrodes.

2. Give the overall operating reaction of the sensor.

3. **a.** Derive the expression for the emf ΔE_T at the sensor terminals as a function of the chemical potentials of the active species and of the CO_2 partial pressure.

   b. Show that we can write it in the form

$$\Delta E_T = \Delta E_T^\circ + f(T, \ln P_{CO_2})$$

   and relate ΔE_T° to the standard free enthalpy $\Delta_r G_T^\circ$ of the overall reaction.

4. The standard free enthalpy $\Delta_r G_T^\circ$ as a function of absolute temperature T of the overall reaction is

$$\Delta_r G_T^\circ = -72\,939 + 149.2\,T \qquad\qquad [J\,mol^{-1}]$$

   a. Using this relationship, derive the relationship $\Delta E_T^\circ = f(T)$.

   b. Compare the results for $\Delta E_{500\,°C}^\circ$ at 500 °C obtained from the experimental data given in figure 88 and from the relationship obtained from the thermodynamic data of question 4(a).

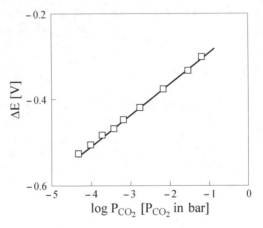

Figure 88 – Sensor potential as a function
of CO_2 partial pressure (from Baliteau, 2005).

   c. Compare the slope of the experimental curve $\Delta E_T = \Delta E_T^\circ + f(\ln P_{CO_2})$ at 500 °C with that obtained from the thermodynamic data at the same temperature and draw your conclusions.

Exercise 5.9 – Sulfur oxide sensor

To analyze the traces of sulfur oxide in gas that is sufficiently rich in oxygen O_2, we associate two solid electrolytes:

▷ the first is an oxide ion O^{2-} (YSZ) conductor,

▷ the second is a silver ion Ag^+ (AgCl) conductor.

The intermediate zone separating the two electrolytes contains the following mix of powders (by volume): 0.5 of metallic silver Ag, 0.25 of the niobium oxide Nb_2O_4, and 0.25 of the niobium oxide Nb_2O_5. We assume that this mix makes a very good electronic conductor and obtain the following electrochemical chain:

$$Pt,O_2 / YSZ / Nb_2O_4 + Nb_2O_5 + Ag / AgCl / Ag_2SO_4 + Pt,SO_3,O_2$$
$$\text{(reference)} \qquad\qquad\qquad\qquad\qquad \text{(measurement)}$$

The operating temperature of the chain is fixed at 500 °C. The two electrodes are in the same dry gas. We assume that, at this temperature,

▷ the sulfur is entirely in the form of SO_3 at the measurement electrode (Mes) and

▷ the reference electrode (Ref) is not sensitive to SO_3.

1. Derive the literal expression for the emf ΔE delivered to the terminals of this chain as a function of the thermodynamic quantities related to the reactions involving Nb_2O_5 and Ag_2SO_4 (see data below). Comment on the particularities of this sensor.

2. Determine
   a. the emf for $P_{SO_3,Mes} = 10^{-7}$ bar,
   b. the variation in the emf when $P_{SO_3,Mes}$ goes to 2×10^{-6} bar.

3. Given that the pellets of solid electrolyte are identically sized with 5 mm diameter and 1 mm thickness, calculate the resistance of the device for analyzing air ($P_{O_2,Mes} = 0.21$ bar). Neglect the resistance of the interfaces.

4. Measurements of the emf at the sensor terminals are made by using a voltmeter with a 1 MΩ input impedance.
   a. Qualitatively, what displacement of substance could this sensor produce?
   b. For these conditions, calculate the lifetime of each consumable substance in the sensor given a 1-mm-thick powder layer mix and that the deposited Ag_2SO_4 covers 1/10 of the surface area with a 10-μm-thick coating.
   c. Which material limits the lifetime?
   d. Which advantage permits the use of a voltage shifter?

Data

$$\Delta_r G_T^\circ(Nb_2O_4 + \tfrac{1}{2}O_2 \longrightarrow Nb_2O_5) = -286\,330 + 64.8\,T\,[J\,mol^{-1}]$$
$$\Delta_r G_T^\circ(2Ag + SO_3 + \tfrac{1}{2}O_2 \longrightarrow Ag_2SO_4) = -336\,800 + 279\,T\,[J\,mol^{-1}]$$

Conductivities $[S\,cm^{-1}]$

▷ Oxygen ion conductor $\sigma_{O^{2-}} = 100\,e^{-\frac{0.8\,eV}{kT}}$

and $\sigma_e = \sigma_0 P_{O_2}^{-\frac{1}{4}}$ with $\sigma_0 = 10^6\,e^{-\frac{3.5\,eV}{kT}}$

▷ Silver ion conductor $\sigma_{Ag^+} = 500\,e^{-\frac{0.4\,eV}{kT}}$ and $\sigma_e = 0$

Element	O	S	Nb	Ag
Atomic mass $[g\,mol^{-1}]$	16	32.1	92.9	107.9

Compound	Ag_2SO_4	Nb_2O_4	Nb_2O_5
Density $[g\,cm^{-3}]$	5.5	5.9	4.5

Exercise 5.10 – Oxygen semiconductor sensor

The conducting properties of titanium dioxide TiO_2 are exploited to make an oxygen sensor. TiO_2 is a non-stoichiometric oxide that, depending on the experimental conditions, contains a metal excess or an oxygen deficiency.

1. For each case,
   a. Propose a formula, describe the phase, and specify the dominant structure elements and electronic defects in the Kröger-Vink notation.
   b. Write the equilibrium reaction between the oxide and gaseous oxygen.
   c. Give the equation that describes the electrical conductivity as a function of oxygen partial pressure, assuming constant electrical mobility.

2. Figure 89 shows the electrical conductivity of titanium oxide as a function of oxygen partial pressure and for several temperatures.

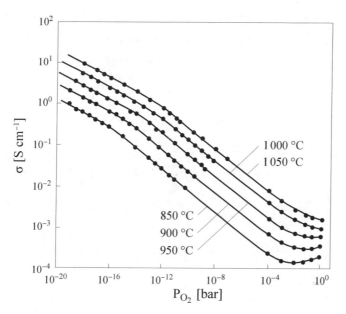

Figure 89 – Electrical conductivity of titanium dioxide
as a function of oxygen partial pressure at various temperatures.

a. Determine the empirical expression $\sigma_e = f(P_{O_2})$ at 1 000 °C
   ▷ in the domain $10^{-12} \leq P_{O_2} \leq 10^{-4}$ bar,
   ▷ in the domain $10^{-20} \leq P_{O_2} \leq 10^{-12}$ bar.
b. Deduce the nature of the dominant point defects in the domain
   $10^{-20} \leq P_{O_2} \leq 10^{-12}$ bar.

3. a. Show that we obtain the equation for conductivity with $P_{O_2}^{-\frac{1}{4}}$ if we consider
      $Ti_i^{3\bullet}$ as majority defect in the oxide instead of $Ti_i^{4\bullet}$.
   b. If we consider oxygen defects as the dominant defects in the oxide, what
      should its effective charge be to obtain an equation for conductivity in
      terms of $P_{O_2}^{-\frac{1}{4}}$?

4. Based on the data given in figure 89 for $P_{O_2} = 10^{-18}$ bar,
   a. determine and tabulate the conductivity of TiO_2 as a function of $1/T$ and
      show that the conductivity of titanium dioxide is activated according to

$$\sigma = \sigma_0 e^{-\frac{E_a}{RT}}$$

   and plot the curve $\quad \log \sigma = f\left(\frac{1}{T}\right)$
   b. determine the pre-exponential factor σ_0 and the activation energy E_a in
      $kJmol^{-1}$.

5. Titanium oxide TiO_2 was used as active material in an oxygen sensor, mainly to regulate the exhaust gas. Based on the properties discussed in questions 2 and (3), describe the operation of this oxygen sensor by stating the main advantages and disadvantages related to its use.

Exercise 5.11 – Amperometric oxygen sensor

1. State the operating principle of an amperometric oxygen sensor. Describe the form of the current-voltage curve for the case of an O_2-Ar gaseous mixture.

2. Propose a detection scheme in which the oxygen supply is limited by a small hole.

3. How is the limiting current I_ℓ related to the molar fraction of oxygen in the gaseous mixture?
 a. in the case of normal diffusion,
 b. in the case of Knudsen diffusion?

4. Table 49 gives the data for the response at 400 °C of an amperometric oxygen sensor for O_2-N_2 mixtures.

Table 49 – Response of amperometric oxygen sensor for O_2-N_2 mixtures at 400 °C.

Molar fraction of oxygen, x_{O_2} [%]	21	40	50	60	80	90
Diffusion-limited current I_ℓ [µA] for 1 bar total pressure	99.7	216	293	388	681	974
Diffusion-limited current I_ℓ [µA] for 1.33×10^{-3} bar total pressure	23.3	47.7	63.3	79.4	111.1	122.2

 a. Plot the curve for the limiting current I_ℓ
  ▷ as a function of the molar fraction x_{O_2} and of $\ln(1-x_{O_2})$ for the case of 1 bar total pressure,
  ▷ as a function of the molar fraction x_{O_2} for the case of 1.33×10^{-3} bar total pressure.
  Give the expression for the function $I_\ell = f(x_{O_2})$ or $I_\ell = f[\ln(1-x_{O_2})]$ based on the results of the preceding question.
 b. What can we conclude?

5. Table 50 presents the data for the limiting current as a function of temperature for an oxygen content of 0.9 and 1 bar total pressure.

Table 50 – Limiting current as a function of temperature
for a molar fraction of oxygen of 0.9 and 1 bar total pressure.

Temperature [°C]	400	500	600	700
Limiting current [µA]	974	1073	1169	1282

Based on the data of table 50,

a. plot the curve representing the limiting current as a function of temperature.

b. Deduce the expression for the function $I_\ell = f(T)$.

6. Describe the shape of the curve for the case of a gaseous mixture containing oxygen and water vapor.

7. What technological improvement have we brought to a hole-based sensor?

Exercise 5.12 – Coulometric oxygen sensor

1. Based on the schematic of a coulometric oxygen sensor shown figure 90, determine the operating mode for this type of sensor.

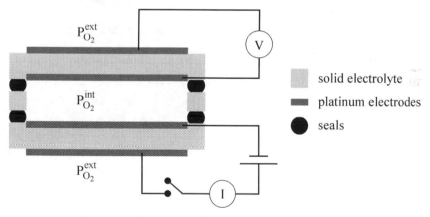

Figure 90 – Schematic of a coulometric oxygen sensor.

2. Determine the oxygen partial pressure in a gaseous mixture of O_2-N_2 by using the response of a coulometric sensor with an interior volume of 0.35 cm^3 operating at 700 °C, given that 9.68 mC of charge is required to equalize the pressures.

3. Other methods have determined with precision the oxygen partial pressure of the gas under analysis to be 0.5 bar. Calculate the percent measurement error due to the use of the coulometric sensor.

4. State the main advantages and disadvantages of solid-electrolyte coulometric gas sensors.

Exercise 5.13 – Nitrogen oxide sensor

We consider a nitrogen oxide sensor made of a nernstian potentiometric cell.

A – Type (I) cell involving sodium ionic conducting electrolyte (NASICON or Na-β-alumina)

To measure NO_2 concentration, chain (Ia) was proposed

$$\text{Na} \quad / \quad \text{NASICON} \quad / \quad \text{NaNO}_3 \quad / \quad \text{Au,NO}_2\text{,O}_2 \qquad \text{(Ia)}$$
$$\text{(reference)} \quad (\text{Na}^+) \quad \text{(porous layer)} \quad \text{(measurement)}$$

The following electrode reaction is considered:

$$\text{Na}^+ + NO_2 + \tfrac{1}{2}O_2 + e \longrightarrow \text{NaNO}_3$$

1. Establish the expression for the emf of the cell (Ia) as a function of
   a. NO_2 partial pressure,
   b. NO_2 molar fraction for the case where total pressure is 1 bar.
   We assume that all thermodynamic equilibria are in place.

2. Figure 91 shows the emf of the cell as a function of NO_2 concentration at 150 °C. Verify that the slope is consistent with the theoretical expression for the emf.

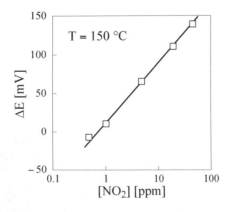

Figure 91 – emf of type (Ia) cell as a function of NO_2 concentration at 150 °C.

3. Calculate the standard free enthalpy of the proposed electrode reaction within the framework of an analysis of air polluted by NO_2 at 150 °C.
Data
$$P_{O_2} = 0.21 \text{ bar}$$

4. To analyze NO, which is a major nitrogen compound in exhaust gas, we propose using a layer of sodium nitrite $NaNO_2$ with the following electrochemical chain:

$$\text{Na} \quad / \quad \text{NASICON} \quad / \quad \text{NaNO}_2 \quad / \quad \text{Au,NO,O}_2 \qquad \text{(Ib)}$$
$$\text{(reference)} \quad (\text{Na}^+) \quad \text{(porous layer)} \text{(measurement)}$$

Give the expression for the emf of the cell (Ib) as a function of NO partial pressure.

5. By using the data of figure 92, determine the number of electrons exchanged in the electrode reaction that we write.

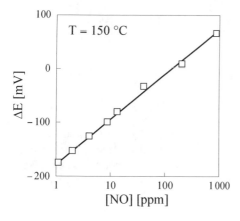

Figure 92 – emf of type (Ib) cell as a function of NO concentration at 150 °C.

6. We observe in figure 93 that the responses of cells (Ia) and (Ib) are independent of oxygen content for a given constant concentration of NO_2 or NO. This result is not consistent with the theoretical relationships developed in questions 1 and 3.

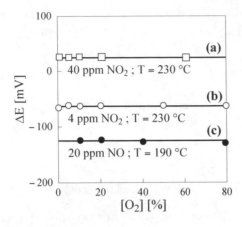

Figure 93 – emf as a function of O$_2$ content in the presence of **(a)** 40 ppm and **(b)** 4 ppm of NO$_2$, cell (Ia) at 230 °C, **(c)** NO cell (Ib) at 190 °C.

Figure 94 shows the emf of these two sensors as a function of oxygen content and in total absence of nitrous oxide.

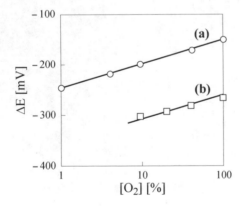

Figure 94 – emf as a function of O$_2$ content and in total absence of nitrous oxide for **(a)** cell (Ia) at 230 °C, **(b)** cell (Ib) at 190 °C.

   a. Calculate the slopes of the lines.

   b. Deduce the number of electrons exchanged in the electrode reaction.

   c. Propose an electrode reaction compatible with the number of electrons exchanged.

7. Based on the experimental results from the preceding questions, propose electrode reactions compatible with the responses for cells (Ia) and (Ib).

8. In the presence of only nitrogen dioxide NO$_2$, we observe that

   ▷ the emf of cell (Ib) corresponds to the exchange of a single electron,

   ▷ the emf is independent of the oxygen partial pressure.

Write an electrode reaction that is simpler than that proposed in question 7 and that accounts for the preceding observations.

9. In sensors (Ia) and (Ib), the reference electrode is formed by the pair Na^+/Na. Its practical implementation is delicate because of the metallic sodium (very reductive, liquid state, cost). Propose another reference electrode that avoids such inconveniences.

B – Type (II) cell involving solid O^{2-} ionic conducting electrolyte $(ZrO_2\text{-}Y_2O_3, ZrO_2\text{-}MgO)$

To measure the concentration of nitrogen oxides at high temperatures and in aggressive conditions (and in particular to analyze automobile exhaust gas), we use sensors such as the following type (IIa) cells:

$$O_2,NO,Pt \ / \ ZrO_2\text{-}Y_2O_3 \ / \ Ba(NO_3)_2 \ / \ Au,NO,O_2 \qquad \text{(IIa)}$$
$$\text{(reference)} \qquad\qquad\qquad\qquad \text{(measurement)}$$

The barium nitrate layer and gold form a porous cermet that allows gaseous species, and in particular nitrogen oxides, to be in direct contact with the stabilized zirconia. Figure 95 shows a schematic representation of the cell.

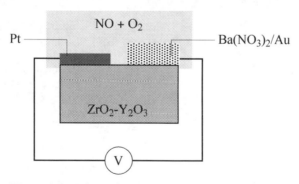

Figure 95 – Schematic representation of type (IIa) cell.

1. Write the electrochemical reactions that occur at each interface.

2. Show that the emf ΔE of the cell may be expressed in the form

$$\Delta E = A + \frac{RT}{4F} \ln P_{O_2} + \frac{RT}{3F} \ln P_{NO}$$

where A is a constant.

Exercise 5.14 – The sodium-sulfur battery

A – Solid electrolyte of sodium-sulfur (Na-S) battery

The solid electrolyte used in a sodium-sulfur battery belongs to the family of sodium β alumina.

1. Give the chemical formula for sodium β alumina.

2. Figure 96 shows the structure of sodium β alumina in perspective view and in the [001] projection. Specify the majority ionic conducting species and the structural characteristic that leads to the high ionic conductivity in this family of electrolytes.

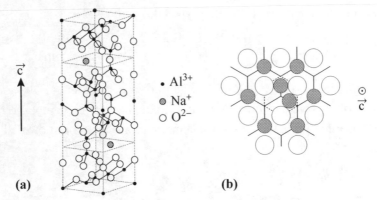

Figure 96 – Representation of **(a)** the unit cell of sodium β alumina (from Kennedy, 1977) and **(b)** the [001] projection (from Déportes *et al.*, 1994).

3. Comment on figure 97, which shows the activation energy of the ionic conductivity as a function of ionic radius of the charge carrier. What do we observe in terms of ionic conductivity if we increase the hydrostatic pressure on the crystal along the $\vec{c}$ axis?

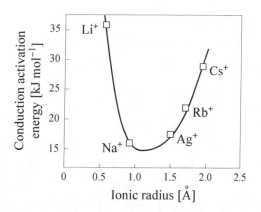

Figure 97 – Activation energy of ionic conductivity as a function of ionic radius of charge carrier (from Whittingham, 1973).

B – Sodium-sulfur battery

The electrolyte for the sodium-sulfur battery is β'' alumina, which has the formula $Na_{1.67}Mg_{0.67}Al_{10.33}O_{17}$.

1. Write the formula in the form $(Na_2O)_\alpha(MgO)_\beta(Al_2O_3)_\gamma$.

2. Starting with $NaNO_3$, $Al(NO_3)_3$, and $Mg(NO_3)_2$, determine the quantities of substances required to make a 300-μm-thick, 20-cm-diameter plate of β'' alumina.

Data

Element	N	O	Na	Mg	Al
Molar mass [g mol^{-1}]	14	16	23	24.3	27

3. Sodium is liquid at the operating temperature of the battery. Give the electrochemical chain that describes the Na-S battery.

4. Write the electrode reactions and the overall operating reaction of the battery.

5. The cathodic reaction occurs in two stages: the first gives polysulfide S_5^{2-}, and the second gives non-stoichiometric polysulfide S_{5-x}^{2-}. Considering only the first stage of reduction, determine the theoretical specific energy W_{th} of the Na-S battery in $W\,h\,kg^{-1}$, given that the potential difference between its terminals is 2.08 V at 300 °C.

6. The technological development of the Na-S battery allows us to obtain a specific energy W_p of nearly 80 $Wh\,kg^{-1}$. Explain this difference between W_{th} and W_p.

7. Propose a simplified scheme for an elementary Na-S battery.

8. Taking $x = 1$ in the polysufide that appears in the second state of the cathodic reaction, evaluate the theoretical capacity Q of charge in a Na-S battery that initially contains 230 g of sodium and 800 g of sulfur. Express Q in Ah.

Exercise 5.15 – General information on fuel cells

1. What is a solid-electrolyte fuel cell?

2. Give three examples of a fuel cell that involves a solid oxide electrolyte. Specify in particular the operating temperature.

3. Gaseous hydrogen H_2 is the simplest fuel to use in fuel cells. Give three examples of processes for manufacturing this fuel.

4. **a.** Describe the shape of the curve for the potential difference and the power density of a fuel cell as a function of the current density that it delivers.
   b. Specify the phenomena that determine the operation of the fuel cell as a function of current density.
   c. At what points $(P_{max}, I_{P_{max}}, \Delta E_{P_{max}})$ is the optimum operation of a fuel cell?

Exercise 5.16 – Solid oxide fuel cell (SOFC)

The following electrochemical chain describes a SOFC:

$$H_2 + H_2O, Ni\text{-}YSZ \;/\; YSZ \;/\; LSM, air$$

where $P_{H_2} = 0.98$ bar, $P_{H_2O} = 0.02$ bar, Ni-YSZ is a ceramic-nickel composite, called "cermet," based on nickel and YSZ, where YSZ is yttria-stabilized zirconia with the formula $(ZrO_2)_{0.92}(Y_2O_3)_{0.08}$, and the cathode LSM is a ceramic with the composition $La_{0.8}Sr_{0.2}MnO_{3-\delta}$.

1. Express the emf ΔE of the cell as a function of the partial pressures of the gases in contact with the electrode.

2. Calculate ΔE at 1 000 K.

Data

> Standard enthalpy of formation of $H_2O_{(g)}$: $\Delta_f H^{\circ}_{1000\,K} = -247.9\,kJ\,mol^{-1}$
> Standard entropy of formation of $H_2O_{(g)}$: $\Delta_f S^{\circ}_{1000\,K} = -55.3\,J\,mol^{-1}\,K^{-1}$
> Percent of oxygen in air: 21%

3. How does the emf ΔE_T vary with temperature when the battery is supplied with the following fuels:

▷ hydrogen H_2,

▷ methane CH_4,

▷ methanol CH_3OH.

4. Determine the thermodynamic efficiency of the cell at 1 000 K in the following cases:

▷ with H_2 as fuel,

▷ total oxidation of CH_4 with oxygen.

Data

> Enthalpy and free molar enthalpy of complete oxidation of the following fuels at 1 000 K:

Fuel	$\Delta_r H$ [kJ mol^{-1}]	$\Delta_r G$ [kJ mol^{-1}]
H_2	-247.9	-192.6
CH_4	-800.5	-800.2

5. Figure 98 shows the ionic conductivity in Arrhenius coordinates of several oxide ion conducting electrolytes.

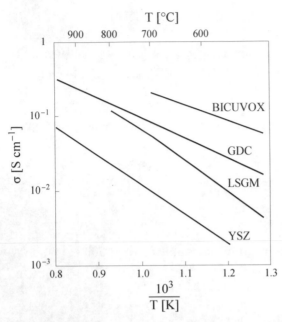

Figure 98 – Ionic conductivity of solid oxide ion conducting electrolytes.

Using this figure, calculate the power density of a SOFC with the following characteristics:

▷ operating temperature: 1 000 K;

▷ electrolyte: YSZ;

▷ diameter of active electrodes: 10 cm;

▷ thickness of electrolyte: $\ell = 0.1$ mm;

▷ cathodic overpotential: $\eta_c = -0.2$ V;

▷ anodic overpotential: $\eta_a = 0.02$ V;

▷ current: I = 23.5 A.

Exercise 5.17 – Use of hydrocarbons in SOFCs

SOFC-type high-temperature fuel cells offer greater flexibility with regards to the fuel compared with low-temperature fuel cells. In addition to hydrogen, carbon monoxide and certain hydrocarbons can be used as fuel, either through chemical reaction in the anodic compartment, or by direct electrochemical oxidation.

The goal of this exercise is to study the different chemical or electrochemical reactions that occur at the electrodes of a SOFC supplied with an arbitrary hydrocarbon fuel C_nH_m.

This exercise does not take into account the kinetic aspects of the reactions.

1. Write the electrode reactions and the overall reaction of the cell assuming hydrogen operation.

2. Chemical oxidation reactions
   a. Write the following reactions assuming methane operation and generalize to the case of an arbitrary hydrocarbon C_nH_m. Apply these reactions to the case of butane C_4H_{10}.
      ▷ water vapor reforming reaction,
      ▷ carbon dioxide reforming reaction,
      ▷ cracking reaction,
      ▷ chemical partial oxidation reaction.
   b. Write the additional reactions that could occur at the anode.

3. Direct electrochemical oxidation reactions
   a. Give the electrode reactions and the overall cell reaction for a SOFC operating on methane in the case of total electrochemical oxidation and in the case of partial electrochemical oxidation (formation of CO and H_2O).
   b. Give the electrode reactions and the overall cell reaction when operating on an arbitrary hydrocarbon fuel C_nH_m for the two following cases and apply these reactions to the case of propane C_3H_8 and but-2-ene C_4H_8:
      ▷ total electrochemical oxidation,
      ▷ partial electrochemical oxidation.
   c. Write the expression for the emf of a cell under total electrochemical oxidation and partial electrochemical oxidation
      ▷ for an arbitrary hydrocarbon C_nH_m,
      ▷ for an alkane with formula C_nH_{2n+2}.

Exercise 5.18 – Thermodynamic study of methane reforming in SOFC

A SOFC can operate directly with methane through various processes, notably internal reforming by water vapor (steam reforming), whose chemical equation is

$$CH_4 + H_2O \rightleftharpoons CO + 3H_2$$

The water-vapor content in the fuel introduced at the anode or, more precisely, the ratio R = S/C (steam S and carbon C) determines the direct internal operation and reforming (for a stoichiometric ratio S/C = 1) or the gradual internal reforming (GIR) (for a stoichiometric ratio S/C < 1). For this exercise, all species are assumed to be in the gas phase.

1. Establish the expression for the emf ΔE of the fuel cell operating on methane.

2. Calculate the standard emf $\Delta E°$ of the pair CH_4/CO_2 at 298 K and at 1 073 K. Comment on how temperature affects $\Delta E°$.

3. Calculate the theoretical emf of the direct internal reforming cell and in GIR for the ratios S/C = ½ and ¼. Consider a residual CO_2 pressure of 4×10^{-3} bar at the anode and an oxygen partial pressure of 0.21 bar at the cathode. At the anode, 10% of the gas introduced consists of a mixture of H_2O/CH_4 in various ratios R, and the remaining 90% is argon. Compare the results with the data from the experimental polarization curves (fig. 99).

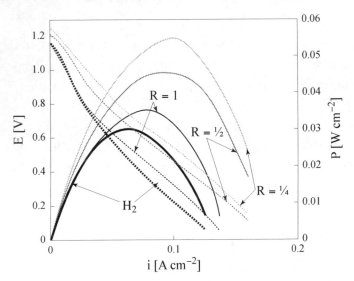

Figure 99 – Polarization and power density curves for SOFC
at 800 °C and for various fuels (from Klein *et al.*, 2009).

4. How does water vapor affect the performance and operation of the cell?

5. Explain how GIR works. Under these conditions, what are the advantages and disadvantages?

Data at 298 K

Compound	$\Delta_f H°$ [kJ mol^{-1}]	$C_p°$ [J mol^{-1}K^{-1}]	$S°$ [J mol^{-1}K^{-1}]
$CH_{4(g)}$	-74.9	23.6	186.2
$O_{2(g)}$	0	29.97	205
$CO_{2(g)}$	-393.5	30.01	213.6
$H_2O_{(g)}$	-241.8	44.16	188.7

Exercise 5.19 – Electrochemical integrator

An electrochemical integrator is a device that measures time simply by reading an emf.

We implement the following all-solid electrochemical chain:

$$C,Ag\text{-}AgCl \: / \: SrCl_2\text{-}NaCl \: / \: C$$
$$(1) \qquad\qquad\qquad (2)$$

The solid solution $SrCl_2$-NaCl is the solid electrolyte where ionic conduction is ensured by chlorine vacancies. Compartment (2) is sealed by a glass bulb under vacuum with a volume of 1 cm^3. The device operates at ambient temperature (T = 298 K).

1. Write the electrode reactions when electrodes (1) and (2) are polarized – and +, respectively.

2. **a.** Given that the electronic conductivity of the electrolyte is negligible and that the gas obeys the ideal gas law, state the relationship between the emf measured after the electric current is stopped, and the integrated current in coulombs that passed through the cell.

   b. What is the emf after the cell carries a 10 mA current for 5 min?

   Data

   | Standard free enthalpy of formation of $AgCl_{(s)}$
   $$\Delta_f G°_{298\,K} = -109.8 \text{ kJ mol}^{-1}$$

   c. Under what conditions can this integrator operate continuously (i.e., the emf may be read directly while the cell carries current)?

   d. What is the maximum measurable current if the mechanical resistance of the bulb is limited to a maximum interior pressure of 2 bar?

Solutions to exercises

Solution 5.1 – Determination of standard free enthalpy of formation for AgCl

1. Expressions for equilibria at the various interfaces

   Interface α $\qquad\qquad e(C_1) \rightleftharpoons e(Ag_{(s)})$ $\qquad\qquad \tilde{\mu}^{\alpha}_{e,Ag(s)} = \tilde{\mu}^{\alpha}_{e,C_1}$

   Interface β $\qquad\qquad Ag_{(s)} \rightleftharpoons Ag^+ + e$ $\qquad \mu^{\beta}_{Ag(s)} = \tilde{\mu}^{\beta}_{Ag^+} + \tilde{\mu}^{\alpha}_{e,Ag(s)}$

   Homogeneous $AgCl_{(s)}$ phase $\quad a^{\beta}_{Ag^+} = a^{\gamma}_{Ag^+}$ $\qquad\qquad \tilde{\mu}^{\beta}_{Ag^+} = \tilde{\mu}^{\gamma}_{Ag^+}$

   Interface γ $\qquad Ag^+ + \frac{1}{2}Cl_{2(g)} + e(C_2) \rightleftharpoons AgCl_{(s)}$

   $$\tilde{\mu}^{\gamma}_{Ag^+} + \frac{1}{2}\mu^{\gamma}_{Cl_2(g)} + \tilde{\mu}^{\gamma}_{e,C_2} = \mu^{\gamma}_{AgCl(s)}$$

   Combining the various equations gives

   $$\Delta E = \frac{\tilde{\mu}^{\alpha}_{e,Cl} - \tilde{\mu}^{\gamma}_{e,C_2}}{F} = \frac{1}{F}\left(\mu^{\beta}_{Ag(s)} + \frac{1}{2}\mu^{\gamma}_{Cl_2(g)} - \mu^{\gamma}_{AgCl(s)}\right)$$

   $$\Delta E = \frac{1}{F}\left(\mu^{\circ\beta}_{Ag(s)} + \frac{1}{2}\mu^{\circ\gamma}_{Cl_2(g)} - \mu^{\circ\gamma}_{AgCl(s)}\right) + \frac{RT}{2F}\ln\frac{P_{Cl_2}}{P^{\circ}}$$

   The formation reaction for $AgCl_{(s)}$ is

   $$Ag_{(s)} + \frac{1}{2}Cl_{2(g)} \rightleftharpoons AgCl_{(s)}$$

   The corresponding standard free enthalpy is

   $$\Delta_f G^{\circ} = \mu^{\gamma}_{AgCl(s)} - \mu^{\circ}_{Ag(s)} - \frac{1}{2}\mu^{\circ}_{Cl_2(g)}$$

   which gives $\qquad\qquad \Delta E = -\frac{\Delta_f G^{\circ}}{F} + \frac{RT}{2F}\ln\frac{P_{Cl_2}}{P^{\circ}}$

   where P° is the standard pressure.

2. The standard free enthalpy of formation of AgCl is

   $$\Delta_f G^{\circ} = -F\Delta E + \frac{RT}{2}\ln\frac{P_{Cl_2}}{P^{\circ}}$$

   $$\Delta_f G^{\circ} = -(0.924 \times 96\,480) + \frac{8.314 \times 623}{2}\ln\frac{5 \times 10^4}{10^5}$$

or $$\Delta_f G^\circ = -90.943 \text{ kJ mol}^{-1}$$

3. The expression for the standard free enthalpy of formation of AgCl in terms of thermodynamic quantities is

$$\Delta_f G^\circ = \Delta_f H^\circ - T\Delta_f S^\circ$$

with $$\Delta_f H^\circ = -127.1 \text{ kJ mol}^{-1}$$

and $$\Delta_f S^\circ = S^\circ(AgCl_{(s)}) - S^\circ(Ag_{(s)}) - \frac{1}{2}S^\circ(Cl_{2(g)})$$

$$\Delta_f S^\circ = 96.2 - 42.5 - \frac{223}{2} = -57.8 \text{ J mol}^{-1} \text{ K}^{-1}$$

We find $$\Delta_f G^\circ = -91.09 \text{ kJ mol}^{-1}$$

which is quite close to the value obtained from the electrochemical cell.

Conclusion – An electrochemical cell may be used to determine precisely the standard enthalpy of formation of certain chemical compounds.

Solution 5.2 – Measurement of thermodynamic quantities of metal fluorides

1. Reaction associated with anionic Frenkel disorder in CaF_2

$$F_F^\times \rightleftharpoons F_i' + V_F^\bullet$$

2. Reaction for doping CaF_2 with
   ▷ NaF $\qquad\qquad$ NaF $\longrightarrow$ $Na_{Ca}' + F_F^\times + V_F^\bullet$
   ▷ YF_3 $\qquad\qquad$ $YF_3 \longrightarrow Y_{Ca}^\bullet + 2F_F^\times + F_i'$

3. Reaction of CaF_2 with water vapor

$$2F_F^\times + H_2O_{(g)} \longrightarrow O_F' + V_F^\bullet + 2HF_{(g)}$$

or $\qquad CaF_2 + H_2O_{(g)} \longrightarrow Ca_{Ca}^\times + O_F' + V_F^\bullet + 2HF_{(g)}$

4. Examples of complexes
   Complexes form from structure defects of opposite effective charge
   ▷ $(Na_{Ca}V_F)^\times$ $\qquad Na_{Ca}' + V_F^\bullet \rightleftharpoons (Na_{Ca}V_F)^\times$ $\qquad$ effective charge $= 0$
   ▷ $(Y_{Ca}F_i)^\times$ $\qquad Y_{Ca}^\bullet + F_i' \rightleftharpoons (Y_{Ca}F_i)^\times$ $\qquad\qquad$ ......

▷ $(O_F V_F)^\times$ $O_F' + V_F^{\bullet\bullet} \rightleftharpoons (O_F V_F)^\times$ ......

▷ $(Y_{Ca} O_F)^\times$ $Y_{Ca}^{\bullet} + O_F' \rightleftharpoons (Y_{Ca} O_F)^\times$ ......

5. The electrochemical chain involved is

$$Pt,Ni\text{-}NiF_2 \,/\, (CaF_2)_{0.99}(NaF)_{0.01} \,/\, M\text{-}MF_2,Pt$$
$$\text{I} \qquad\qquad\qquad\qquad\qquad \text{II}$$

a. The emf ΔE measured at the terminals of the chain is due to the chemical-potential gradient of fluorine between the two electrodes, each of which is in the following oxidation-reduction equilibrium:

$$F_{2(g)} + 2e_{Pt} \rightleftharpoons 2F^-_{(SE)}$$

The emf is $\Delta E = \dfrac{RT}{2F} \ln \dfrac{P_{F_2(I)}}{P_{F_2(II)}}$

where $P_{F_2(I)}$ and $P_{F_2(II)}$ denote the fluorine partial pressures in compartments I and II, respectively.

We thus arrive at $P_{F_2(II)} = P_{F_2(I)}\, e^{-\frac{2F\Delta E}{RT}}$

or $P_{F_2(II)} = 7.4 \times 10^{-39} e^{-\frac{2 \times 96\,480 \times 0.794}{8.314 \times 1000}}$

$$P_{F_2(II)} = 7.35 \times 10^{-47}\ \text{bar}$$

b. The formation reaction for the compound MF_2 is

$$M_{(s)} + F_{2(g)} \longrightarrow MF_{2(s)}$$

and the corresponding standard free enthalpy of formation is

$$\Delta_f G_T^\circ = RT \ln P_{F_2(II)} \qquad\qquad \text{with } P_{F_2(II)} \text{ in bars}$$

At 1 000 K, we obtain

$$\Delta_f G_{1000\,K}^\circ = 8.314 \times 1000 \ln(7.35 \times 10^{-47})$$

$$\Delta_f G_{1000\,K}^\circ = -883.2\ \text{kJ mol}^{-1}$$

c. Comparing with the data in the problem statement shows that the metal used in compartment II is manganese.

Solution 5.3 – Measurement of O^{2-} ion activity in a molten salt

1. The electric potential within the electrochemical chain is shown schematically below (fig. 100).

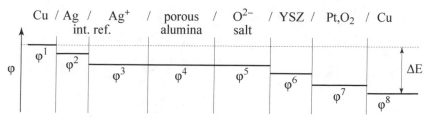

Figure 100 – Qualitative variation of potential within the chain.

The emf is also shown.

2. To determine the emf of the sensor as a function of the concentration of O^{2-} ions dissolved in the molten salt, we write below the equilibrium at each interface and the corresponding equations.

▷ Electronic equilibrium at interface α and in phase (2)

$$\tilde{\mu}_e^{\alpha,Cu} = \tilde{\mu}_e^{\alpha,Ag} = \tilde{\mu}_e^{\beta,Ag}$$

or $\qquad \mu_e^{Cu} - F\varphi^1 = \mu_e^{Ag} - F\varphi^2$

▷ Electrochemical equilibrium at interface β

$$Ag^+ + e \rightleftharpoons Ag$$

$$\tilde{\mu}_{Ag^+}^{int.ref.} + \tilde{\mu}_e^{\beta,Ag} = 0$$

or $\qquad \mu_{Ag^+}^{int.ref.} + F\varphi^3 = -\mu_e^{Ag} + F\varphi^2$

▷ Ionic equilibrium at interface ε and in phase (5)

$$\tilde{\mu}_{O^{2-}}^{\varepsilon,salt} = \tilde{\mu}_{O^{2-}}^{\varepsilon,YSZ} = \tilde{\mu}_{O^{2-}}^{\lambda,YSZ}$$

or $\qquad \mu_{O^{2-}}^{salt} - 2F\varphi^5 = \mu_{O^{2-}}^{YSZ} - 2F\varphi^6$

▷ Electrochemical equilibrium at interface λ

$$\tfrac{1}{2}O_{2,gas} + 2e_{Pt} \rightleftharpoons O_{YSZ}^{2-}$$

$$\tfrac{1}{2}\mu_{O_2} + 2\tilde{\mu}_e^{\lambda,Pt} = \tilde{\mu}_{O^{2-}}^{\lambda,YSZ}$$

or $\qquad \tfrac{1}{2}\mu_{O_2} + 2\mu_e^{Pt} - 2F\varphi^7 = \mu_{O^{2-}}^{YSZ} - 2F\varphi^6$

▷ Electronic equilibrium at interface χ

$$\tilde{\mu}_e^{\chi,Pt} = \tilde{\mu}_e^{\chi,Cu}$$

or
$$\mu_e^{Pt} - F\varphi^7 = \mu_e^{Cu} - F\varphi^8$$

Combining the preceding relations allows us to write

$$\Delta E = \varphi^8 - \varphi^1 = \frac{1}{F}\left[-\mu_{Ag^+}^{int.ref.} - \frac{1}{2}\mu_{O^{2-}}^{salt} + \frac{1}{4}\mu_{O_2}\right]$$

Because
$$\mu_{Ag^+}^{int.ref.} = \mu_{Ag^+}^\circ + RT\ln a_{Ag^+}^{int.ref.}$$

$$\mu_{O^{2-}}^{salt} = \mu_{O^{2-}}^\circ + RT\ln a_{O^{2-}}^{salt}$$

$$\mu_{O_2} = \mu_{O_2}^\circ + RT\ln P_{O_2}$$

given that
$$P_{O_2} = 0.2 \text{ bar}$$

By inserting this into the expression for the emf ΔE, we obtain

$$\Delta E = -\frac{1}{F}\left[\mu_{Ag^+}^\circ + RT\ln a_{Ag^+}^{int.ref.} + \frac{1}{2}\mu_{O^{2-}}^\circ + \frac{RT}{2}\ln a_{O^{2-}}^{salt} - \frac{RT}{4}\ln P_{O_2}\right]$$

By joining activity and molality in molten salts, we obtain

$$\mu_{Ag^+}^{int.ref.} = \mu_{Ag^+}^\circ + RT\ln m_{Ag^+}^{int.ref.}$$

$$\mu_{O^{2-}}^{salt} = \mu_{O^{2-}}^\circ + RT\ln m_{O^{2-}}^{salt}$$

or
$$\mu_{O^{2-}}^{salt} = \mu_{O^{2-}}^\circ - 2.3\,RT\,pO^{2-}$$

and $\Delta E = -\frac{1}{F}\left[\mu_{Ag^+}^\circ + RT\ln m_{Ag^+}^{int.ref.} + \frac{1}{2}\mu_{O^{2-}}^\circ - \frac{RT}{4}\ln 0.2\right] - \frac{2.3RT}{2F}\log m_{O^{2-}}^{salt}$

By writing $\quad \Delta E^\circ = -\frac{1}{F}\left[\mu_{Ag^+}^\circ + RT\ln m_{Ag^+}^{int.ref.} + \frac{1}{2}\mu_{O^{2-}}^\circ - \frac{RT}{4}\ln 0.2\right]$

we have
$$\Delta E = \Delta E^\circ - \frac{2.3RT}{2F}\log m_{O^{2-}}^{salt}$$

and
$$\Delta E = \Delta E^\circ + \frac{2.3RT}{2F}pO^{2-}$$

3. **a.** For high pO^{2-}, the solution is not saturated, so the emf varies with pO^{2-}. The horizontal part corresponds to a saturated solution. The intersection between the two line segments indicates the onset of oxide precipitation and allows the solubility to be evaluated.

   b. Determination of experimental slope

$$S_{expt} = \frac{\Delta(\Delta E)}{\Delta pO^{2-}}$$

for the case of ZnO, we obtain

$$s_{expt} \approx 95\,mV\,u_{pO^{2-}}^{-1}$$

where $u_{pO^{2-}}^{-1}$ denotes the units of pO^{2-}.

The result of question 2 allows us to calculate the theoretical slope s_{th}.

$$s_{th} = \frac{2.3RT}{2F}$$

$$s_{th} = \frac{2.3 \times 8.314 \times 723}{2 \times 96\,480}$$

$$s_{th} = 71.7\,mV\,u_{pO^{2-}}^{-1}$$

We observe that $s_{expt} \approx s_{th}$, which validates the expression for the emf as a function of pO^{2-} established in question 2.

c. The solubility equilibria and the corresponding constants
   ▷ for ZnO are $ZnO_{(s)} \rightleftharpoons Zn^{2+} + O^{2-}$

   with $K_s(ZnO) = [Zn^{2+}]\,[O^{2-}]$

                         $s(ZnO) = [O^{2-}]$

   where s denotes the solubility.

   ▷ for Li_2O are $Li_2O_{(s)} \rightleftharpoons 2Li^+ + O^{2-}$

   with $K_s(Li_2O) = [Li^+]^2[O^{2-}]$

   and $s(Li_2O) = [O^{2-}]$

d. From figure 83 the solubilities of ZnO and Li_2O may be read from the respective intersections marking the onset of precipitation.
   ▷ For ZnO $\log s(ZnO) = pO_{sat}^{2-} = -1.8$

                    $s = 1.58 \times 10^{-2}\,mol\,kg^{-1}$

   ▷ For Li_2O $\log s(Li_2O) = pO_{sat}^{2-} = -1.95$

                    $s = 1.12 \times 10^{-2}\,mol\,kg^{-1}$

Solution 5.4 – Calculation of equilibrium constants for defect formation in Cu_2O

1. ▷ Formation of singly ionized copper vacancies

$$\tfrac{1}{4}O_{2(g)} \rightleftharpoons \tfrac{1}{2}O_O^X + V_{Cu}' + h^\bullet$$

with
$$K_{V'_{Cu}} = \frac{[V'_{Cu}] \times [h^\bullet]}{P_{O_2}^{1/4}}$$

▷ Formation of neutral copper vacancies

Neutral copper vacancies result from an equilibrium between ionized copper vacancies and holes by the reaction

$$V_{Cu}^\times \rightleftharpoons V'_{Cu} + h^\bullet$$

with
$$K_{V_{Cu}} = \frac{[V'_{Cu}] \times [h^\bullet]}{[V_{Cu}^\times]}$$

from which the equilibrium for the formation of neutral copper vacancies is

$$\tfrac{1}{4}O_{2(g)} \rightleftharpoons \tfrac{1}{2}O_O^\times + V_{Cu}^\times$$

with
$$K_{V_{Cu}^\times} = \frac{[V_{Cu}^\times]}{P_{O_2}^{1/4}}$$

and
$$K_{V'_{Cu}} = K_{V_{Cu}^\times} \times K_{V_{Cu}}$$

▷ Formation of interstitial singly ionized oxygen

$$\tfrac{1}{2}O_{2(g)} \rightleftharpoons O'_i + h^\bullet$$

with
$$K_{O'_i} = \frac{[O'_i] \times [h^\bullet]}{P_{O_2}^{1/2}}$$

2. The expressions for the defect concentrations are deduced from the preceding relationships:

$$[V'_{Cu}] = \frac{K_{V'_{Cu}} \times P_{O_2}^{1/4}}{[h^\bullet]} \quad \text{and} \quad [O'_i] = \frac{K_{O'_i} \times P_{O_2}^{1/2}}{[h^\bullet]}$$

Enforcing the condition of electroneutrality gives

$$[h^\bullet] = [V'_{Cu}] + [O'_i]$$

from which
$$[h^\bullet] = \sqrt{\left(K_{V'_{Cu}} \times P_{O_2}^{1/4}\right) + \left(K_{O'_i} \times P_{O_2}^{1/2}\right)}$$

Moreover, the p-type electronic conductivity is expressed as

$$\sigma_h = u_h F [h^\bullet]$$

where u_h is electrical mobility of holes.

By using
$$k_i = (u_h F)^2 K_i$$

with i = V'_{Cu} or O'_i we finally obtain

$$\sigma = \sqrt{\left(k_{V'_{Cu}} \times P_{O_2}^{\frac{1}{4}}\right) + \left(k_{O'_i} \times P_{O_2}^{\frac{1}{2}}\right)}$$

This expression is the theoretical equation for Cu_2O conductivity as a function of oxygen partial pressure based on the defect model established in question 1.

3. **a.** Fitting this theoretical equation by least squares to the experimental points allows us to determine the parameters in the expression above for conductivity. We obtain

▷ at 1 000 °C $\sigma = \sqrt{\left(42.39 \times P_{O_2}^{\frac{1}{4}}\right) + \left(85.14 \times P_{O_2}^{\frac{1}{2}}\right)}$

▷ at 1 100 °C $\sigma = \sqrt{\left(135.46 \times P_{O_2}^{\frac{1}{4}}\right) + \left(247.01 \times P_{O_2}^{\frac{1}{2}}\right)}$

b. The evolution of the Cu_2O conductivity as a function of oxygen partial pressure at 1 000 and 1 100 °C is shown on a logarithmic scale in figure 101.

The curves reproduce fairly well the experimental data.

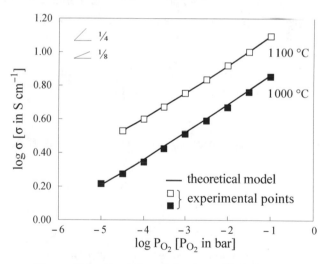

Figure 101 – Evolution of Cu_2O conductivity as a function of oxygen partial pressure at 1 000 and 1 100 °C.

4. Considering a hole mobility of $1\ m^2V^{-1}s^{-1}$, the coefficients k determined from the fits allow us to determine $K_{V'_{Cu}}$ and $K_{O'_i}$ with $K = \dfrac{k}{(u_h F)^2}$.

a. ▷ at 1 000 °C $K_{V'_{Cu}} = \dfrac{42.39}{96\,480^2}$

▷ at 1 100 °C $K_{V'_{Cu}} = \dfrac{135.46}{96\,480^2}$

We obtain $K_{V'_{Cu}}^{1000\,°C} = 4.55 \times 10^{-9}$

and $K_{V'_{Cu}}^{1100\,°C} = 1.45 \times 10^{-8}$

b. ▷ at 1 000 °C $K_{O'_i} = \dfrac{85.14}{96\,480^2}$

▷ at 1 100 °C $K_{O'_i} = \dfrac{247.01}{96\,480^2}$

We obtain $K_{O'_i}^{1000\,°C} = 9.15 \times 10^{-9}$

and $K_{O'_i}^{1100\,°C} = 2.65 \times 10^{-8}$

Solution 5.5 – TiS$_2$: insertion material

1. In electrochemistry, an insertion material is a solid phase, crystalline or amorphous, capable of hosting and allowing the displacement of ionic and electronic species. This type of phase makes it possible to insert and extract ionic species, and in particular alkaline metals.

2. Here are some examples of insertion materials for lithium that are used in primary and secondary batteries: MnO_2, V_6O_{13}, V_2O_5, $NiPS_3$, MoS_3, $FePO_3$, ...

3. The following three conditions must be satisfied to insert lithium into an insertion material:

   ▷ the existence of vacant sites capable of hosting a Li^+ ion,

   ▷ the existence of an electronic energy band containing free levels (unfilled orbitals) that can be occupied by electrons,

   ▷ conservation of the material integrity during the insertion in order to maintain a fixed number of hosting sites for the lithium.

4. **a.** The electrochemical reactions are

   ▷ at the anode $xLi \rightleftarrows xLi^+ + xe$

   ▷ at the cathode $TiS_2 + xe \rightleftarrows TiS_2^{x-}$

   ▷ the overall operating reaction being

$$xLi + TiS_2 \rightleftarrows Li_x TiS_2$$

b. In the cathode composed of $Li_x TiS_2$, we have xLi^+, xTi^{3+}, and $(1-x)Ti^{4+}$. The formula may be written in the form $Li_x^+ Ti_x^{3+} Ti_{1-x}^{4+} S_2$.

▷ This equation may be obtained under the following conditions:
- the integrity of the TiS_2 structure is conserved while the Li is being inserted, which means that the number of ionic sites remains constant;
- we assume as a first approximation that the unoccupied sites are equivalent and that their energy is independent of the insertion fraction;
- we consider that the fill factor of the TiS_2 electronic band remains relatively constant, which implies a large number of electronic states;
- this electronic band must be populated before insertion to ensure permanent electronic conduction.

▷ Based on the thermodynamic model, for $x = \frac{1}{2}$

$$E^{\frac{1}{2}} = E^\circ$$

from which we get $\qquad E^\circ = 2.215\ V$

Consequently, the equation is

$$E_{Li} = 2.215 - 2.57 \times 10^{-2} \ln \frac{x}{1-x}$$

▷ Curve showing experimental open-circuit emf as a function of insertion fraction (fig. 102)

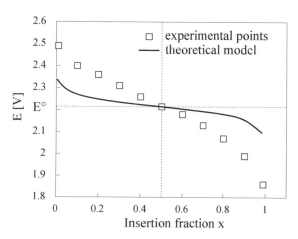

Figure 102 – Electric potential as a function of insertion fraction.

We see that the agreement between theory and the experimental points is not satisfactory.

c. ▷ Use of the equation

$$E_{Li} = E° - \frac{RT}{F} \ln \frac{x}{x_m - x} - z \frac{E_{int}}{F} \frac{x}{x_m}$$

Based on the theoretical model, for $x = \frac{1}{2}$

$$E° = E^{\frac{1}{2}} + \frac{z}{2F} E_{int}$$

The equation can be reformulated as follows:

$$E_{Li} = 2.215 - \frac{E_{int}}{2 \times 96\,480}(3 - 6x) - 2.57 \times 10^2 \ln \frac{x}{1 - x}$$

▷ A least squares fit to the experimental points leads to a result for E_{int} that agrees better with the experimental points

$$E_{int} = 6.11 \, kJ \, mol^{-1}$$

Reasonable agreement is found between the experimental points and the theoretical model (see fig. 103).

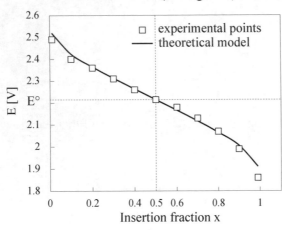

Figure 103 – Electric potential as a function of insertion fraction. The theoretical model accounts for the interaction energy.

5. The TiS_2 groups form compact lamellar layers connected by weak Van der Waals bonds. The inserted Li^+ ions position themselves between the layers, accompanied by a slight expansion of the crystal lattice (fig. 104). Lithium and titanium occupy octahedral sites in a compact sulfur sublattice.

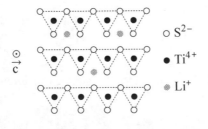

o S^{2-}

● Ti^{4+}

⊛ Li^+

Figure 104 – Structural representation of Li_xTiS_2 for [001] projection.

The relative change in volume, ΔV [%], is given by the expression

$$\Delta V[\%] = \frac{V(LiTiS_2) - V(TiS_2)}{V(TiS_2)} \times 100\%$$

We can choose a volume element V corresponding to a prism with a triangular base (equilateral) with side a and height c

$$V = \frac{a^2 \sqrt{3}}{2} \times c$$

Numerical evaluation gives $V(LiTiS_2) = 64.04$ Å^3 and $V(TiS_2) = 57.26$ Å^3, which leads to the following relative change in volume:

$$\Delta V = 11.8\%$$

Solution 5.6 – Chlorine sensor based on doped strontium chloride

A – Study of strontium chloride $SrCl_2$

1. **a.** The structure defects related to anionic Frenkel disorder in $SrCl_2$ are chlorine vacancies $V_{Cl}^{\bullet}$ and interstitial chlorine Cl_i'. The corresponding equilibrium is $\qquad Cl_{Cl}^{\times} \rightleftharpoons V_{Cl}^{\bullet} + Cl_i'$
   and the corresponding constant K_{AF} is given by

   $$K_{AF} = [V_{Cl}^{\bullet}] \times [Cl_i']$$

   b. The various temperature domains shown in figure 85(a) are identified as follows:
   ▷ The BC domain, which corresponds to the smallest slope, represents the extrinsic domain; in other words, the domain where the concentration of the charge carrier that gives rise to electrical conduction is fixed by foreign species (impurities). It dominates over the concentration of intrinsic species. Here, the Cl^- ion migrates *via* a vacancy mechanism.
   ▷ The AB domain corresponds to the dominant concentration of intrinsic species $V_{Cl}^{\bullet}$ and Cl'_i. The conduction is implemented by the most mobile species. Here again, the Cl^- ion migrates *via* a vacancy mechanism.
   ▷ The CD domain, which appears at low temperatures, is the zone where we observe a strong interaction between effective-charge defects of opposite sign with pair formation, "clusters," or the precipitation of new phases. The decrease in conductivity is due to the decrease in concentration of the most mobile charge carrier.

c. The mobilities will be equal when

$$\frac{6 \times 10^{-3}}{T} e^{-\frac{0.4\,eV}{kT}} = \frac{36}{T} e^{-\frac{1\,eV}{kT}}$$

or

$$T = \frac{0.6\,eV}{k \times \ln(6 \times 10^3)}$$

$$T = \frac{0.6 \times 1.6 \times 10^{-19} \times 6.02 \times 10^{23}}{8.314 \times \ln(6 \times 10^3)}$$

$$T = 799\,K$$

d. ▷ Given that the conductivity of particle i is given by

$$\sigma_i = z_i^2 F^2 \tilde{u}_i [i]$$

the total conductivity of $SrCl_2$ under the conditions of question (c) is

$$\sigma_t = \sigma_{V_{Cl}^{\bullet}} + \sigma_{Cl_i'}$$

$$\sigma_t = F^2 \tilde{u}_{Cl_i'} ([V_{Cl}^{\bullet}] + [Cl_i'])$$

By neglecting the electronic conductivity, the electroneutrality equation takes the form $[Cl_i'] = [V_{Cl}^{\bullet}] = \sqrt{K_{AF}}$

The final expression for the total conductivity is thus

$$\sigma_t = 2 F^2 \tilde{u}_{Cl_i'} \sqrt{K_{AF}}$$

▷ Numerical evaluation gives

$$\sigma_t = 2 \times (96\,480)^2 \times \frac{36}{799} e^{-\frac{1\,eV}{kT}} \times 3 \times 10^{-2} e^{-\frac{0.8\,eV}{kT}}$$

$$\sigma_t = 2 \times (96\,480)^2 \times \frac{36}{799} e^{-\frac{1.6 \times 10^{-19}}{1.38 \times 10^{-23} \times 799}} \times 3 \times 10^{-2} e^{-\frac{0.8 \times 1.6 \times 10^{-19}}{1.38 \times 10^{-23} \times 799}}$$

$$\sigma_t = 1.14 \times 10^{-4}\,S\,cm^{-1}$$

2. a. The reaction for dissolution of KCl in $SrCl_2$ is

$$KCl \longrightarrow K_{Sr}' + Cl_{Cl}^{\times} + V_{Cl}^{\bullet}$$

It leads to an increase in the concentration of chlorine vacancies $V_{Cl}^{\bullet}$.

b. In figure 85(b),

▷ the $B'C'$ domain corresponding to the smallest slope represents the extrinsic domain; that is, the domain where the concentration of charge carriers that give rise to electrical conduction is fixed by foreign species. In this case, this involves essentially the dopant KCl.

This concentration dominates over that of species of intrinsic origin.

▷ the C′D′ domain is interpreted in the same manner as the CD domain in question 1(b).

3. a. Figure 105 shows the Brouwer diagrams $\log [i] = f(\log P_{Cl_2})$, where i denotes a structural defect. The doping translates into an increase in $[V_{Cl}^{\bullet}]$ and the widening of the electrolytic domain.

b. The use of the solid solution $SrCl_2$-KCl as electrolyte is justified by the presence of a large ionic domain, where the concentration of chlorine vacancies of extrinsic origin is high and constant because of the dopant KCl.

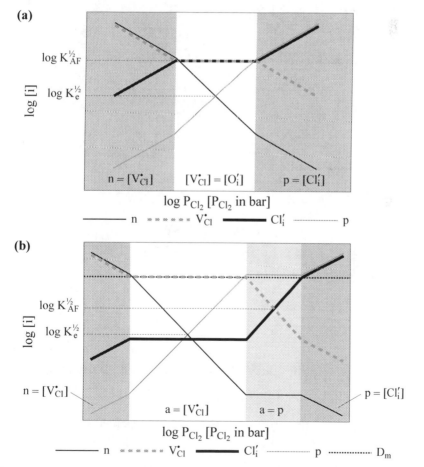

Figure 105 – Brouwer diagram for **(a)** pure $SrCl_2$ and **(b)** solid solution $SrCl_2$-KCl.

4. a. The reaction for doping the solid solution $SrCl_2$-KCl by $AgCl$ is

$$AgCl + V^{\bullet}_{Cl} \longrightarrow Ag^{\bullet}_i + Cl^{\times}_{Cl}$$

b. After doping with $AgCl$, the electroneutrality equation is (neglecting electronic disorder) $[Ag^{\bullet}_i] + [V^{\bullet}_{Cl}] = [Cl'_i] + [K'_{Sr}]$

By using $[V^{\bullet}_{Cl}] = x$, we obtain

$$b + x = \frac{K_{AF}}{x} + a$$

Because $a = b$, we have $x^2 = K_{AF}$

or $x = [V^{\bullet}_{Cl}] = \sqrt{K_{AF}}$ and $Cl'_i = \dfrac{K_{AF}}{\sqrt{K_{AF}}} = \sqrt{K_{AF}}$

The expression for ionic conductivity is

$$\sigma'_t = \sigma_{V^{\bullet}_{Cl}} + \sigma_{Cl'_i} + \sigma_{Ag^{\bullet}_i}$$

In the framework of question 1, we have

$$\sigma'_t = F^2 \tilde{u}_{Cl'_i}([V^{\bullet}_{Cl}] + [Cl'_i]) + F^2 \tilde{u}_{Ag^{\bullet}_i}[Ag^{\bullet}_i]$$

$$\sigma'_t = \sigma_t + F^2 \tilde{u}_{Ag^{\bullet}_i}[Ag^{\bullet}_i]$$

where σ_t is the total ionic conductivity of pure $SrCl_2$ expressed in question 1.

c. Numerical evaluation gives

$$\sigma'_t = 1.14 \times 10^{-4} + 96\,480^2 \times \frac{6 \times 10^2}{799} e^{-\frac{1.6 \times 10^{-19}}{1.38 \times 10^{-23} \times 799}} \times 3 \times 10^{-2} e^{-\frac{0.8 \times 1.6 \times 10^{-19}}{1.38 \times 10^{-23} \times 799}}$$

$$\sigma'_t = 1.06 \times 10^{-3} \; S\,cm^{-1}$$

B – *Study of chlorine sensor*

1. We consider the following electrochemical chain:

$$Cl_2(P_1),Me \; / \; SrCl_2\text{-}KCl \; / \; Me,Cl_2(P_2)$$
$$(1) \hspace{4cm} (2)$$

The two electrodes host the following electrochemical reaction:

$$\tfrac{1}{2}Cl_{2(g)} + e_{Me} \rightleftarrows Cl^-_{SE}$$

SE refers to the solid electrolyte $SrCl_2$-KCl, which conducts by chloride ions Cl^-.

2. To establish the expression for the emf ΔE at the terminals of the electrochemical chain, we write the equilibria at the interfaces

▷ Compartment (1) $\frac{1}{2}Cl_{2(g)} + e_{Me} \rightleftharpoons Cl_{SE}^-$

with $\frac{1}{2}\mu_{Cl_2}^{(1)} + \tilde{\mu}_e^{Me1} = \tilde{\mu}_{Cl^-}^{SE1}$

Cl^- is the majority carrier in the solid electrolyte

$$\tilde{\mu}_{Cl^-}^{SE1} = \tilde{\mu}_{Cl^-}^{SE2}$$

▷ Compartment (2) $\frac{1}{2}Cl_{2(g)} + e_{Me} \rightleftharpoons Cl_{SE}^-$

with $\frac{1}{2}\mu_{Cl_2}^{(2)} + \tilde{\mu}_e^{Me2} = \tilde{\mu}_{Cl^-}^{SE2}$

Combining the three preceding equations gives

$$\frac{1}{2}\mu_{Cl_2}^{(1)} + \tilde{\mu}_e^{Me1} = \frac{1}{2}\mu_{Cl_2}^{(2)} + \tilde{\mu}_e^{Me2}$$

$$\frac{1}{2}\mu_{Cl_2}^{\circ} + \frac{RT}{2}\ln P_{Cl_2}^{(1)} + \mu_e^{Me1} - F\varphi^{(1)} = \frac{1}{2}\mu_{Cl_2}^{\circ} + \frac{RT}{2}\ln P_{Cl_2}^{(2)} + \mu_e^{Me2} - F\varphi^{(2)}$$

$$F(\varphi^{(1)} - \varphi^{(2)}) = \frac{RT}{2}\ln \frac{P_{Cl_2}^{(1)}}{P_{Cl_2}^{(2)}}$$

By using $\Delta E = \varphi^{(1)} - \varphi^{(2)}$, we arrive at

$$\Delta E = \frac{RT}{2F}\ln \frac{P_{Cl_2}^{(1)}}{P_{Cl_2}^{(2)}}$$

This is the expression for the emf as a function of chlorine partial pressures.

3. **a.** The electrochemical chain to consider in this case is

$$Ag\text{-}AgCl\,/\,SrCl_2\text{-}KCl\,/\,Ag,Cl_2$$

We present the various equilibria in the chain

$$AgCl_{(s)} + e_{Ag} \rightleftharpoons Ag_{(s)} + Cl_{SE}^-$$

with $\mu_{AgCl} + \tilde{\mu}_e^{Ag1} = \mu_{Ag} + \tilde{\mu}_{Cl^-}^{SE1}$

Cl^- being the majority carrier in the solid electrolyte, we have

$$\tilde{\mu}_{Cl^-}^{SE1} = \tilde{\mu}_{Cl^-}^{SE2}$$

(1) and (2) denote the internal and external faces of the electrolyte.

$$\frac{1}{2}Cl_{2(g)} + e_{Ag} \rightleftharpoons Cl_{SE,2}^-$$

with $\frac{1}{2}\mu_{Cl_2} + \tilde{\mu}_e^{Ag2} = \tilde{\mu}_{Cl^-}^{SE2}$

Combining these reactions gives

$$\mu_{AgCl} + \tilde{\mu}_e^{Ag1} = \frac{1}{2}\mu_{Cl_2} + \tilde{\mu}_e^{Ag2} + \mu_{Ag}$$

$$\mu_{AgCl} + \mu_e^{Ag1} - F\varphi^{(1)} = \tfrac{1}{2}\mu_{Cl_2} + \mu_e^{Ag2} - F\varphi^{(2)} + \mu_{Ag}$$

$$F(\varphi^{(2)} - \varphi^{(1)}) = \tfrac{1}{2}\mu_{Cl_2} + \mu_{Ag} - \mu_{AgCl}$$

Because $Ag_{(s)}$ and $AgCl_{(s)}$ form distinct solid phases, we obtain

$$F(\varphi^{(2)} - \varphi^{(1)}) = \tfrac{1}{2}\mu_{Cl_2}^{\circ} + \tfrac{RT}{2}\ln P_{Cl_2} + \mu_{Ag}^{\circ} - \mu_{AgCl}^{\circ}$$

The emf $\Delta E = \varphi^{(2)} - \varphi^{(1)}$ at the chain terminals is expressed as

$$\Delta E = \tfrac{1}{F}\left[\mu_{Ag}^{\circ} + \tfrac{1}{2}\mu_{Cl_2}^{\circ} - \mu_{AgCl}^{\circ}\right]\tfrac{RT}{2F}\ln P_{Cl_2}$$

with P_{Cl_2} in bars.

b. The mixture Ag-AgCl plays the role of reference electrode.

4. a. Figure 106 shows the emf as a function of absolute temperature. We observe that this function is linear and is given by

$$\Delta E = -0.6\,T + 1331\;[mV]$$

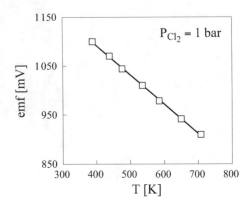

Figure 106 – Electromotive force of sensor as a function of temperature.

b. Under the measurement conditions ($P_{Cl_2} = 1$ bar), the expression for the emf is

$$\Delta E = \tfrac{1}{F}\left(\mu_{Ag}^{\circ} + \tfrac{1}{2}\mu_{Cl_2}^{\circ} - \mu_{AgCl}^{\circ}\right)$$

The second member of this equation is equal to the opposite of the change $\Delta_f G^{\circ}$ in the standard free enthalpy of formation of $AgCl_{(s)}$:

$$\Delta E = -\frac{\Delta_f G_{AgCl_{(s)}}^{\circ}}{F}$$

Ellingham approximation gives

$$\Delta E = -\frac{\Delta_f H_{AgCl_{(s)}}^{\circ} - T\Delta_f S_{AgCl_{(s)}}^{\circ}}{F}$$

where $\Delta_f H^\circ_{AgCl_{(s)}}$ and $\Delta_f S^\circ_{AgCl_{(s)}}$ are respectively the change in the standard enthalpy and in the standard entropy of formation of $AgCl_{(s)}$. This relationship indicates that the emf is a linear function of temperature under these measurement conditions, which we verify in figure 106.

c. The change in the free enthalpy of formation of AgCl is given by

$$\Delta_f G^\circ_{AgCl_{(s)}} = -\Delta E \times F$$

Numerical evaluation at 350 °C gives

$$\Delta E = -0.6 \times 623 + 1\,331$$

$$\Delta E = 957.2 \text{ mV}$$

from which $\Delta_f G^\circ_{623\,K,AgCl_{(s)}} = -957.2 \times 10^{-3} \times 96\,480$

$$\Delta_f G^\circ_{623\,K,AgCl_{(s)}} = -92.4 \text{ kJ mol}^{-1}$$

Solution 5.7 – CO_2 sensor (a)

1. **a.** The electrochemical chain is

$$\text{Au1,Na / SE / Na}_2\text{CO}_3\ (\text{O}_{2(g)},\text{CO}_{2(g)}),\text{Au2}$$
$$\quad\ \alpha\quad\ \beta\qquad \gamma$$

b. Analytic expression of the emf at the sensor terminals
We consider the following equilibria:

▷ at interface α $\qquad 2e_{Au1} \rightleftharpoons 2e_{Na}$

$$2\tilde{\mu}_e^{Au1} = 2\tilde{\mu}_e^{Na}$$

$$2\mu_e^{Au1} - 2F\varphi_{Au1} = 2\mu_e^{Na} - 2F\varphi_{Na}$$

▷ at interface β $\qquad 2Na \rightleftharpoons 2Na^+ + 2e$

$$2\mu_{Na}^{Na} = 2\tilde{\mu}_{Na^+}^{SE} + 2\mu_e^{Na} - 2F\varphi_{Na}$$

▷ at interface γ $\qquad 2Na_{SE}^+ \rightleftharpoons 2Na_{Na_2CO_3}^+$

$$2\tilde{\mu}_{Na^+}^{SE} = 2\tilde{\mu}_{Na^+}^{Na_2CO_3} = 2\tilde{\mu}_{Na^+}$$

and $2Na^+ + CO_2 + \frac{1}{2}O_2 + 2e \rightleftharpoons Na_2CO_3$

$$\mu_{Na_2CO_3} = 2\tilde{\mu}_{Na^+} + \mu_{CO_2} + \frac{1}{2}\mu_{O_2} + 2\mu_e^{Au2} - 2F\varphi_{Au2}$$

Combining the preceding relationships allows us to express the emf ΔE:

$$\Delta E = \varphi_{Au2} - \varphi_{Au1}$$

$$\Delta E = \frac{1}{2F}\left(2\mu_{Na} + \mu_{CO_2} + \tfrac{1}{2}\mu_{O_2} - \mu_{Na_2CO_3}\right)$$

The term in parentheses is the opposite of the change $\Delta_r G_T$ in the free enthalpy of the following reaction:

$$2Na + CO_2 + \tfrac{1}{2}O_2 \rightleftharpoons Na_2CO_3$$

which gives

$$\Delta E = -\frac{\Delta_r G_T}{2F}$$

$$\Delta E = \frac{1}{2F}\left(-\Delta_r G_T^\circ + RT \ln P_{CO_2} + \frac{RT}{2}\ln P_{O_2}\right)$$

where $\Delta_r G_T^\circ$ is the change in the standard free enthalpy of reaction (1) and P_{CO_2} and P_{O_2} are the partial pressures of CO_2 and O_2, respectively, expressed in bars.

2. Numerical evaluation gives

$$\Delta E = \frac{1}{2 \times 96\,480}\left[742\,370 - (276 \times 473) + (8.314 \times 473 \times \ln 10^3)\right.$$
$$\left. + \left(\frac{8.314 \times 473}{2} \times \ln 0.21\right)\right]$$

$$E = 3.014 \text{ V}$$

3. The potential difference between the sensor terminals for $P_{CO_2} = 10^{-5}$ bar is

$$\Delta E = \frac{1}{2 \times 96\,480}\left[742\,370 - (276 \times 473) + (8.314 \times 473 \times \ln 10^{-5})\right.$$
$$\left. + \left(\frac{8.314 \times 473}{2} \times \ln 0.21\right)\right]$$

$$\Delta E = 2.92 \text{ V}$$

The small input impedance R of the voltmeter used in the measurement causes a current I that ceases when all the sodium of the reference electrode is consumed.

The current I is given by

$$I = \frac{\Delta E}{R}$$

$$I = \frac{2.92}{10^7} = 2.92 \times 10^{-7} \text{ A}$$

The lifetime t of the sensor is given by the charge Q that passes through the circuit,

$$Q = I \times t = n \times F$$

where n is the number of moles of sodium contained in the reference electrode.

We have
$$n = \frac{m}{M} = \frac{\rho V}{M}$$

where ρ, M, and V are respectively the density, molar mass, and volume of the reference electrode. The lifetime is given by
$$t = \frac{n \times F}{I} = \frac{\rho \times V \times F}{I \times M}$$

Numerical evaluation gives
$$t = \frac{0.97 \times 0.4 \times 10^{-4} \times 96\,480}{2.92 \times 10^{-7} \times 23}$$

$$t = 5.57 \times 10^5 \text{ s}$$

$$t = 155\,h$$

4. Figure 107 shows, for three temperatures, the change in the theoretical emf of the sensor as a function of the logarithm of the CO_2 partial pressure. The lines intersect at the isoelectric point P_i, where the emf is independent of temperature.

A linear regression allows us to determine the theoretical equation governing the change in the emf as a function of CO_2 partial pressure

▷ at 298 K $\Delta E = 2.95 \times 10^{-2} \log P_{CO_2} + 3.40$

▷ at 353 K $\Delta E = 3.49 \times 10^{-2} \log P_{CO_2} + 3.32$

We deduce the CO_2 partial pressure at the isoelectric point of the sensor
$$(3.49 - 2.95) \times 10^{-2} \log P_{CO_2} = 3.40 - 3.32$$

$$\log P_{CO_2} = \frac{3.40 - 3.32}{(3.49 - 2.95) \times 10^{-2}}$$

$$\log P_{CO_2} = 14.81$$

and
$$P_{CO_2} = 6.45 \times 10^{14} \text{ bar}$$

This result is far from the pressure of CO_2 in the ambient air.

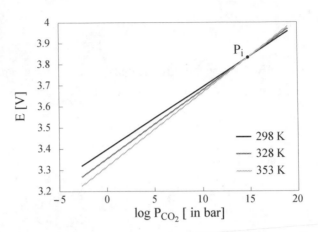

Figure 107 – Change in theoretical emf of the sensor as a function
of the logarithm of the CO_2 partial pressure for three temperatures.

5. a. The electrochemical chain of the sensor with the new reference electrode
is $Au1, Ag / AgBr\text{-}NaBr / SE / Na_2CO_3, O_{2(g)}, CO_{2(g)} / Au2$.

b. To determine the analytic expression of the emf ΔE at the sensor terminals, we consider the following equilibria:

$$2e_{Au1} \rightleftharpoons 2e_{Ag}$$

$$2e_{Ag} + 2Ag^+_{Ref} \rightleftharpoons 2Ag_{Ref}$$

$$2AgBr_{Ref} \rightleftharpoons 2Ag^+_{Ref} + 2Br^-_{Ref}$$

$$2Br^-_{Ref} + 2Na^+_{Ref} \rightleftharpoons 2NaBr_{Ref}$$

$$2Na^+_{SE} \rightleftharpoons 2Na^+_{Ref}$$

$$2Na^+_W \rightleftharpoons 2Na^+_{SE}$$

$$Na_2CO_{3,W} \rightleftharpoons 2Na^+_W + CO_2 + \tfrac{1}{2}O_2 + 2e_W$$

$$2e_W \rightleftharpoons 2e_{Au2}$$

Adding these equilibrium reactions together term by term gives

$$2e_{Au1} + 2AgBr_{Ref} + Na_2CO_3 \rightleftharpoons 2Ag_{Ref} + 2NaBr_{Ref} + CO_2 + \tfrac{1}{2}O_2 + 2e_{Au2}$$

and $2(\tilde{\mu}_e^{Au1} - \tilde{\mu}_e^{Au2}) = 2\mu_{Ag} + 2\mu_{NaBr} + \mu_{CO_2} + \tfrac{1}{2}\mu_{O_2} - \mu_{Na_2CO_3} - 2\mu_{AgBr}$

$$2F(\varphi_{Au2} - \varphi_{Au1}) = 2\mu_{Ag} + 2\mu_{NaBr} + \mu_{CO_2} + \tfrac{1}{2}\mu_{O_2} - \mu_{Na_2CO_3} - 2\mu_{AgBr}$$

By using $\Delta E = \varphi_{Au2} - \varphi_{Au1}$
we obtain

$$2\,F\Delta E \;=\; 2\mu^{\circ}_{NaBr} - 2\mu^{\circ}_{AgBr} + 2RT\ln\frac{a_{NaBr}}{a_{AgBr}} - \mu^{\circ}_{Na_2CO_3}$$

$$+\;\mu^{\circ}_{CO_2} + RT\ln P_{CO_2} + \frac{RT}{2}\ln P_{O_2}$$

where a_{NaBr} and a_{AgBr} denote respectively the activities of sodium bromide and silver bromide in the mixture NaBr-AgBr. If we fix a_{NaBr} and a_{AgBr}, the analytical expression for the emf at temperature T is

$$\Delta E \;=\; A + \frac{RT}{2F}\ln P_{CO_2} + \frac{RT}{4F}\ln P_{O_2}$$

with

$$A \;=\; \frac{\mu^{\circ}_{NaBr}}{F} - \frac{\mu^{\circ}_{AgBr}}{F} + \frac{RT}{F}\ln\frac{a_{NaBr}}{a_{AgBr}} - \frac{\mu^{\circ}_{Na_2CO_3}}{2F} - \frac{\mu^{\circ}_{CO_2}}{2F} \;=\; \text{const.}$$

Note – To operate this CO_2 sensor, the oxygen partial pressure must be held constant.

c. The impedance of the Ag/AgBr-NaBr reference, which is much greater than that of the metallic sodium reference, cannot be neglected. The use of a low-input-impedance voltmeter leads to a smaller current than observed for the sensor studied in question 2.

Solution 5.8 – CO$_2$ sensor (b)

1. Presentation of existing equilibria

   ▷ at the reference electrode

   $$Na_2Ti_6O_{13} \;\rightleftharpoons\; 6TiO_2 + \tfrac{1}{2}O_2 + 2Na^+_{NASICON} + 2e_{RE} \qquad (1)$$

   ▷ at the measurement electrode

   $$CO_2 + \tfrac{1}{2}O_2 + 2Na^+_{NASICON} + 2e_{ME} \;\rightleftharpoons\; Na_2CO_3 \qquad (2)$$

   The species Na^+ is in equilibrium in the phases $Na_2Ti_6O_{13}$, NASICON, and Na_2CO_3.

2. The overall reaction describing the sensor operation is

   $$Na_2CO_3 + 6TiO_2 \;\rightleftharpoons\; Na_2Ti_6O_{13} + CO_2$$

3. a. Expression of the emf ΔE_T at the sensor terminals as a function of the chemical potentials of the active species and of the CO_2 partial pressure

▷ At the reference electrode (1), we can write

$$6\mu^{\circ}_{TiO_2} + \frac{1}{2}\mu^{\circ}_{O_2} + \frac{RT}{2}\ln\frac{P_{O_2}}{P^{\circ}} + 2\tilde{\mu}^{NASICON}_{Na^+} + 2\tilde{\mu}^{RE}_e = \mu^{\circ}_{Na_2Ti_6O_{13}}$$

▷ and at the measurement electrode (2), we can write

$$\mu^{\circ}_{CO_2} + RT\ln\frac{P_{CO_2}}{P^{\circ}} + \frac{1}{2}\mu^{\circ}_{O_2} + \frac{RT}{2}\ln\frac{P_{O_2}}{P^{\circ}}$$
$$+ 2\tilde{\mu}^{NASICON}_{Na^+} + 2\tilde{\mu}^{ME}_e = \mu^{\circ}_{Na_2CO_3}$$

where P° is the standard pressure. By developing the chemical potential of electrons, we obtain

$$6\mu^{\circ}_{TiO_2} + \frac{1}{2}\mu^{\circ}_{O_2} + \frac{RT}{2}\ln\frac{P_{O_2}}{P^{\circ}} + 2\tilde{\mu}^{NASICON}_{Na^+}$$
$$+ 2\mu^{RE}_e - 2F\varphi_{RE} = \mu^{\circ}_{Na_2Ti_6O_{13}}$$

and

$$\mu^{\circ}_{CO_2} + RT\ln\frac{P_{CO_2}}{P^{\circ}} + \frac{1}{2}\mu^{\circ}_{O_2} + \frac{RT}{2}\ln\frac{P_{O_2}}{P^{\circ}}$$
$$+ 2\tilde{\mu}^{NASICON}_{Na^+} + 2\mu^{ME}_e - 2F^{ME}_\varphi = \mu^{\circ}_{Na_2CO_3}$$

Finally, we deduce the emf ΔE_T

$$\Delta E_T = \varphi_{ME} - \varphi_{RE}$$
$$= \frac{1}{2F}\left(\mu^{\circ}_{Na_2Ti_6O_{13}} + \mu^{\circ}_{CO_2} - 6\mu^{\circ}_{TiO_2} - \mu^{\circ}_{Na_2CO_3}\right) + \frac{RT}{2F}\ln\frac{P_{CO_2}}{P^{\circ}}$$

b. The expression $\left(\mu^{\circ}_{Na_2Ti_6O_{13}} + \mu^{\circ}_{CO_2} - 6\mu^{\circ}_{TiO_2} - \mu^{\circ}_{Na_2CO_3}\right)$ is the opposite of the change in the standard free enthalpy of the operating reaction

$$-\Delta_r G^{\circ}_T = \mu^{\circ}_{Na_2Ti_6O_{13}} + \mu^{\circ}_{CO_2} - 6\mu^{\circ}_{TiO_2} - \mu^{\circ}_{Na_2CO_3}$$

By using $$\Delta E^{\circ}_T = -\frac{\Delta_r G^{\circ}_T}{2F}$$

we obtain $$\Delta E_T = \Delta E^{\circ}_T + \frac{RT}{2F}\ln\frac{P_{CO_2}}{P^{\circ}}$$

4. a. Expression for $\Delta E^{\circ}_T = f(T)$

Given that $\Delta_r G^{\circ}_T = -72\,939 + 149.2\,T\ [J\ mol^{-1}]$

we write $$\Delta E^{\circ}_T = -\frac{\Delta_r G^{\circ}_T}{2F}$$

$$\Delta E^{\circ}_T = -\frac{72\,939 + 149{,}2\,T}{2 \times 96\,480}$$

$$\Delta E_T^\circ = -0.378 - 7.73 \times 10^{-4}\,T\ [V]$$

b. Use of the thermodynamic data gives

$$\Delta E_{500\,°C}^\circ = -0.378 - (7.73 \times 10^{-4} \times 773) = -0.220\ V$$

Exploitation of the experimental curve in figure 88 leads to

$$\Delta E_{500\,°C}^\circ = -0.217\ V$$

The difference is close to 1.4%. We find that the thermodynamic data satisfactorily account for the sensor response. This agreement validates the choice of equilibria considered in question 1.

c. The characteristic slope of the experimental curve is $s_{expt} = 0.078$ V per decade. The slope of the equation describing the thermodynamic data is given by

$$s_{calc} = \frac{RT}{2F}\ln 10$$

$$s_{calc} = \frac{8.314 \times 773}{2 \times 96\,480}\ln 10$$

$$s_{calc} = 0.077\ V\ \text{per decade}$$

The difference between the two slopes is of the order of 1.28%. These results indicate that the sensor response is nernstian and confirms the good agreement between the thermodynamic data and experiment.

Solution 5.9 – Sulfur oxide sensor

1. The potential ΔE between the terminals of the electrochemical chain is

$$\Delta E = \varphi_{Mes} - \varphi_{Ref}$$

where φ_{Mes} and φ_{Ref} are the potentials of the measurement and reference electrodes. ΔE may equally well be written in the form

$$\Delta E = \varphi_{Mes} - \varphi_{Ag} + \varphi_{Ag} - \varphi_{Ref}$$

where φ_{Ag} is the potential of the metallic silver in the chain.

▷ Let us first calculate $\varphi_{Mes} - \varphi_{Ag}$. For this, we write
  • the equilibrium at the measurement electrode,

$$SO_{3(g)} + 2Ag^+ + 2e_{Pt}^- + \tfrac{1}{2}O_{2(g)} \rightleftharpoons Ag_2SO_{4(s)}$$

which gives

$$\mu_{SO_3}^{\circ} + RT \ln P_{SO_3}^{Mes} + 2\tilde{\mu}_{Ag^+} + 2\mu_e$$

$$- 2F\varphi_{Mes} + \frac{1}{2}\mu_{O_2}^{\circ} + \frac{RT}{2} \ln P_{O_2}^{Mes} = \mu_{Ag_2SO_4}^{\circ}$$

with P expressed in bars.

- the internal equilibrium relative to the Ag^+/Ag couple,

$$2Ag^+ + 2e_{Ag} \;\rightleftharpoons\; 2Ag$$

from which we obtain $2\tilde{\mu}_{Ag^+} + 2\mu_e - 2F\varphi_{Ag} = 2\mu_{Ag}^{\circ}$

Combining the two relations deduced from the two equilibria gives

$$\varphi_{Mes} - \varphi_{Ag} = \frac{1}{2F}\left(2\mu_{Ag}^{\circ} - \mu_{Ag_2SO_4}^{\circ} + \mu_{SO_3}^{\circ} + \frac{1}{2}\mu_{O_2}^{\circ}\right)$$

$$+ \frac{RT}{2F} \ln P_{SO_3}^{Mes} + \frac{RT}{4F} \ln P_{O_2}^{Mes}$$

$$\varphi_{Mes} - \varphi_{Ag} = \frac{-\Delta_r G_{Ag_2SO_4}^{\circ}}{2F} + \frac{RT}{2F} \ln P_{SO_3}^{Mes} + \frac{RT}{4F} \ln P_{O_2}^{Mes}$$

▷ We then calculate $\varphi_{Ag} - \varphi_{Ref}$. The equilibria to consider are

$$Nb_2O_4 + O^{2-} \;\rightleftharpoons\; Nb_2O_5 + 2e$$

with $\mu_{Nb_2O_4}^{\circ} + \tilde{\mu}_{O^{2-}} = \mu_{Nb_2O_5}^{\circ} + 2\mu_e - 2F\varphi_{Ag}$

and $\frac{1}{2}O_{2(g)} + 2e_{Ref} \;\rightleftharpoons\; O_{Ref}^{2-}$

with $\frac{1}{2}\mu_{O_2}^{\circ} + \frac{RT}{2} \ln P_{O_2}^{Ref} + 2\mu_e^{Ref} - 2F\varphi_{Ref} = \tilde{\mu}_{O^{2-}}^{Ref}$

Assuming that the equilibria of electronic species and of the O^{2-} species are established between the different phases at the reference electrode, we obtain

$$\varphi_{Ag} - \varphi_{Ref} = \frac{1}{2F}\left(\mu_{Nb_2O_5}^{\circ} - \mu_{Nb_2O_4}^{\circ} - \frac{1}{2}\mu_{O_2}^{\circ}\right) - \frac{RT}{4F} \ln P_{O_2}^{Ref}$$

$$\varphi_{Ag} - \varphi_{Ref} = \frac{\Delta_r G_{Nb_2O_5}^{\circ}}{2F} - \frac{RT}{4F} \ln P_{O_2}^{Ref}$$

Given that the oxygen partial pressure is the same at the two electrodes, we obtain the expression for the potential ΔE between the chain terminals:

$$\Delta E = \varphi_{Mes} - \varphi_{Ref} = \frac{\Delta_r G_{Nb_2O_5}^{\circ} - \Delta_r G_{Ag_2SO_4}^{\circ}}{2F} + \frac{RT}{2F} \ln P_{SO_3}^{Mes}$$

This chain thus constitutes a $SO_{3(g)}$ sensor. It operates as a second species sensor because it requires Ag_2SO_4. The presence of two membranes allows us to eliminate the oxygen partial pressure.

2. Numerical evaluation

   a. Value of the potential for $P_{SO_3}^{Mes} = 10^{-7}$ bar at 500 °C

   $$\Delta E = \frac{-286\,330 + (64.8 \times 773) + 336\,800 - (279 \times 773)}{2 \times 96\,480}$$
   $$+ \frac{8.14 \times 773}{2 \times 96\,480} \ln 10^{-7}$$

   $$\Delta E = -1.133\ V$$

   b. Change in potential when $P_{SO_3}^{Mes}$ reaches 2×10^{-6} bar at 500 °C

   $$\delta(\Delta E) = \frac{RT}{2F} \ln \frac{(P_{SO_3}^{Mes})_2}{(P_{SO_3}^{Mes})_1}$$

   $$\delta(\Delta E) = \frac{8.14 \times 773}{2 \times 96\,480} \ln \frac{2 \times 10^{-6}}{10^{-7}}$$

   $$\delta(\Delta E) = 0.099\ V$$

3. The resistance R of the device is equal to the sum of the resistances R_o of the oxide ion conducting electrolyte, denoted o, and R_{Ag} of the silver-ion conducting electrolyte, denoted Ag,

   $$R = R_o + R_{Ag}$$

   $$R = \frac{1}{\sigma_o}\left(\frac{\ell}{S}\right)_o + \frac{1}{\sigma_{Ag}}\left(\frac{\ell}{S}\right)_{Ag}$$

   Because the pellets have the same geometric factor

   $$k = \left(\frac{\ell}{S}\right)_o = \left(\frac{\ell}{S}\right)_{Ag}$$

   we have

   $$R = k\left(\frac{1}{\sigma_o} + \frac{1}{\sigma_{Ag}}\right)$$

   Determination of the conductivities of both electrolytes

   ▷ Oxide ion conducting electrolyte

   $$\sigma_o = \sigma_{O^{2-}} + \sigma_e$$

   $$\sigma_o = 100\,e^{-\frac{0.8\,eV}{kT}} + 10^6 P_{O_2}^{-\frac{1}{4}} e^{-\frac{3.5\,eV}{kT}}$$

$$\sigma_o = 100\,e^{-\frac{0.8\times1.6\times10^{-19}}{1.38\times10^{-23}\times773}} + 10^6 \times 0.21^{-1/4}\,e^{-\frac{3.5\times1.6\times10^{-19}}{1.38\times10^{-23}\times773}}$$

$$\sigma_o = 6.14\times10^{-4} + 2.34\times10^{-17}\,\mathrm{S\,cm^{-1}} = 6.14\times10^{-4}\,\mathrm{S\,cm^{-1}}$$

▷ Silver-ion conducting electrolyte

$$\sigma_{Ag} = 500\,e^{-\frac{0.4\,eV}{kT}}$$

$$\sigma_{Ag} = 500\,e^{-\frac{0.4\times1.6\times10^{-19}}{1.38\times10^{-23}\times773}}$$

$$\sigma_{Ag} = 1.24\,\mathrm{S\,cm^{-1}}$$

Numerical evaluation gives

$$R = \frac{0.1\times4}{\pi\times0.5^2}\left(\frac{1}{6.14\times10^{-4}} + \frac{1}{1.24}\right)$$

or
$$R = 830.3\,\Omega$$

4. a. With the sensor connected to the terminals of the voltmeter, an electric current runs through the circuit because the input impedance of the voltmeter is not sufficiently high. The $SO_{3(g)}$ measurement electrode with Ag_2SO_4 is the negative electrode. The electric current inside the sensor circulates from Ag_2SO_4 to the reference oxygen electrode. The result is

▷ the consumption of Ag_2SO_4 and the production of silver in the intermediate zone near the mixture of Nb_2O_5/Nb_2O_4 and Ag^+,

▷ the formation of O^{2-} ions that will react with Nb_2O_4 according to the reaction

$$Nb_2O_5 + 2e \longrightarrow Nb_2O_4 + O^{2-}$$

b. The current I running through the circuit is

$$I = \frac{\Delta E}{R_{imp}}$$

$$I = \frac{1.133}{10^6} = 1.133\times10^{-6}\,\mathrm{A}$$

where R_{imp} is the input impedance of the measurement instrument. The volume of Ag_2SO_4 is

$$V_{Ag_2SO_4} = \frac{\pi\times0.5^2}{4}\times0.1\times10^{-3} = 1.96\times10^{-5}\,\mathrm{cm^3}$$

which corresponds to a mass

$$m_{Ag_2SO_4} = 5.5\times1.96\times10^{-5} = 1.078\times10^{-4}\,\mathrm{g}$$

The molar mass of Ag_2SO_4 is

$$M_{Ag_2SO_4} = (2 \times 107.9) + 32.1 + (4 \times 16) = 311.9 \, g\,mol^{-1}$$

We deduce the number of moles of Ag_2SO_4

$$n_{Ag_2SO_4} = \frac{1.078 \times 10^{-4}}{311.9} = 3.46 \times 10^{-7} \, mol$$

The time t required for all the Ag_2SO_4 to disappear is

$$t = \frac{Q}{I} = \frac{2F \times n_{Ag_2SO_4}}{I}$$

$$t = \frac{2 \times 96\,480 \times 3.46 \times 10^{-7}}{1.133 \times 10^{-6}}$$

$$t = 58\,927\,s = 16.36\,h$$

We repeat the same calculation for Nb_2O_4. The volume of Nb_2O_4 is

$$V_{Nb_2O_4} = \frac{\pi \times 0.5^2}{4} \times 0.1 \times \frac{1}{4} = 4.91 \times 10^{-3} \, cm^3$$

The molar mass of Nb_2O_4 is

$$M_{Nb_2O_4} = (2 \times 92.9) + (4 \times 16) = 249.8 \, g\,mol^{-1}$$

The number of moles of Nb_2O_4 is

$$n_{Nb_2O_4} = \frac{4.91 \times 10^{-3} \times 5,9}{249.8} = 1.16 \times 10^{-4} \, mol$$

The time t' required for all the Nb_2O_4 to disappear is

$$t' = \frac{2F \times n_{Nb_2O_4}}{I}$$

$$t' = \frac{2 \times 96\,480 \times 1.16 \times 10^{-4}}{1.133 \times 10^{-6}}$$

$$t' = 1.97 \times 10^7 \, s = 5\,488\,h$$

c. The results show that the amount of Ag_2SO_4 is what limits the lifetime of the sensor.

d. The use of a voltage shifter allows us to decrease the current I. For example, if we shift the voltage by 1 V, we would have

$$\Delta E' = -1.133 + 1 = -0.133 \, V$$

▷ The current is approximately divided by 10.
▷ We work at a more sensitive scale.

Solution 5.10 – Oxygen semiconductor sensor

1. a. Description of the phase

    ▷ Excess metal: we can propose the formula $Ti_{1+x}O_2$ with

$$Ti_{1+x}O_2 \equiv Ti_{Ti}^x + 2O_O^x + xTi_i^{4\bullet} + 4xe'$$

Electroneutrality gives $4[Ti_i^{4\bullet}] = [e'] = n$

where we have neglected p.

    ▷ Oxygen defects: we can propose the formula $TiO_{2(1-x)}$ with

$$TiO_{2(1-x)} \equiv Ti_{Ti}^x + 2(1-x)\,O_O^x + 2xV_O^{\bullet\bullet} + 4xe'$$

Electroneutrality gives $2[V_O^{\bullet\bullet}] = [e'] = n$

where we have neglected p.

b. Presentation of the equilibrium reaction with the gaseous phase

    ▷ Excess of metal: $Ti_{1+x}O_2$

$$Ti_{Ti}^x + 2O_O^x \rightleftharpoons O_{2(g)} + Ti_i^{4\bullet} + 4e' \tag{1}$$

    ▷ Oxygen defects: $TiO_{2(1-x)}$

$$2O_O^x \rightleftharpoons O_{2(g)} + 2V_O^{\bullet\bullet} + 4e' \tag{2}$$

c. Variation in the electronic conductivity with oxygen partial pressure assuming that the electrical mobility u_e is constant

In both cases, the expression for the electronic conductivity is

$$\sigma_e = F \times u_e \times n$$

The variation of n with oxygen partial pressure P_{O_2} is obtained in each case based on the equilibrium constant K_g of the reaction involving $O_{2(g)}$.

    ▷ Excess metal: $Ti_{1+x}O_2$

$$K_{g1} = [Ti_i^{4\bullet}] \times n^4 \times P_{O_2}$$

By using the electroneutrality relation, we obtain

$$K_{g1} = \frac{1}{4} \times n^5 \times P_{O_2}$$

or $n = A_1\,P_{O_2}^{-\frac{1}{5}}$

with $A_1 = (4K_{g1})^{\frac{1}{5}} = const.$

and $\sigma_{e1} = A_1 \times F \times u_e \times P_{O_2}^{-\frac{1}{5}}$

$$\sigma_{el} = \sigma_1^\circ \times P_{O_2}^{-\frac{1}{5}}$$

where σ_1° is a constant.

▷ Oxygen defect: $TiO_{2(1-x)}$

$$K_{g2} = [V_O^{\cdot\cdot}]^2 \times n^4 \times P_{O_2}$$

By using the electroneutrality equation, we obtain

$$K_{g2} = \frac{1}{4} \times n^6 \times P_{O_2}$$

or

$$n = A_2 P_{O_2}^{-\frac{1}{6}}$$

with

$$A_2 = (4K_{g2})^{\frac{1}{6}} = \text{const.}$$

and

$$\sigma_{e2} = A_2 \times F \times u_e \times P_{O_2}^{-\frac{1}{6}}$$

$$\sigma_{e2} = \sigma_2^\circ \times P_{O_2}^{-\frac{1}{6}}$$

where σ_2° is a constant.

2. **a.** Experimental equation $\log \sigma_e = f(\log P_{O_2})$ at $1\,000\,°C$
    ▷ By exploiting the results shown in figure 89, we obtain in the domain $10^{-12} \le P_{O_2} \le 10^{-4}$ bar an equation of the type

$$\log \sigma = \log \sigma^\circ + \log P_{O_2}^{-\frac{1}{4}}$$

    ▷ Likewise, in the domain $10^{-12} \le P_{O_2} \le 10^{-20}$ bar, we obtain an equation of the type

$$\log \sigma = \log \sigma^\circ + \log P_{O_2}^{-\frac{1}{6}}$$

Remark – The preceding reasoning associates the total conductivity to the electronic conductivity of the semiconductor. This is completely justified because of the differences in concentration and, especially, in the mobility of the ionic and electronic carriers.

  b. Referring to question 1(c), we notice that the majority point defect is the oxygen vacancy $V_O^{\cdot\cdot}$ compensated by the electronic species e'.

3. **a.** We consider the equilibrium with gaseous oxygen in the following reaction:

$$Ti_{Ti}^{\times} + 2O_O^{\times} \rightleftharpoons O_{2(g)} + Ti_i^{3\cdot} + 3e' \qquad (3)$$

The equilibrium constant and the corresponding electroneutrality equation are respectively $\quad K_{g3} = [Ti_i^{3\bullet}] \times n^3 \times P_{O_2}$

$$3[Ti_i^{3\bullet}] = [e'] = n$$

and the combination gives $\quad K_{g3} = \dfrac{1}{3} \times n^4 \times P_{O_2}$

or $\qquad\qquad\qquad\qquad n = A_3 \, P_{O_2}^{-\frac{1}{4}}$

with $\qquad\qquad\qquad A_3 = (3K_{g3})^{\frac{1}{4}} = \text{const.}$

and $\qquad\qquad\qquad \sigma_{e3} = A_3 \times F \times u_e \times P_{O_2}^{-\frac{1}{4}}$

$$\sigma_{e3} = \sigma_3^\circ \times P_{O_2}^{-\frac{1}{4}}$$

where σ_3° is a constant.

By associating the total conductivity with the electronic conductivity, we deduce

$$\log \sigma = \log \sigma_3^\circ + \log P_{O_2}^{-\frac{1}{4}}$$

b. The following equilibrium reaction with gaseous oxygen must be considered:

$$2O_O^X \rightleftharpoons O_{2(g)} + 2V_O^{\bullet\bullet} + 2e' \qquad\qquad (4)$$

The equilibrium constant K_{e4} and the corresponding electroneutrality equation are respectively

$$K_{g4} = [V_O^{\bullet\bullet}]^2 \times n^2 \times P_{O_2}$$

$$[V_O^{\bullet\bullet}] = [e'] = n$$

A reasoning identical to that used in 3(a) leads to the following result:

$$\log \sigma = \log \sigma_4^\circ + \log P_{O_2}^{-\frac{1}{4}}$$

where σ_4° is a constant.

If the majority defect is the oxygen vacancy, its effective charge must be equal to $+1$.

4. a. Based on the numerical data extracted from figure 89 for $P_{O_2} = 10^{-18}$ bar and listed in table 51, we see that the curve shown in figure 108 is linear, which means that the electrical conductivity in TiO_2 is an activated phenomenon.

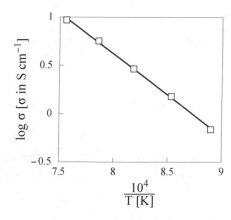

Figure 108 – Logarithm of conductivity of TiO_2 as a function of inverse temperature.

Table 51 – Logarithm of conductivity of TiO_2 as a function of inverse temperature for $P_{O_2} = 10^{-18}$ bar.

$10^4/T\,[K]$	8.9	8.53	8.18	7.86	7.56
$\log \sigma\,[S\,cm^{-1}]$	– 0.164	0.179	0.463	0.752	0.97

b. The pre-exponential term and the activation energy E_a are obtained by linear regression. We find

$$\sigma^\circ = 2.56 \times 10^7\,S\,cm^{-1}$$

and $$E_a = 162.2\,kJ\,mol^{-1}$$

Finally, the expression for conductivity as a function of temperature is

$$\sigma = 2.56 \times 10^7 e^{-\frac{8481}{T}}\,S\,cm^{-1}$$

5. Based on the properties discussed in questions 2 and 3, we conclude that the operating principle of an oxygen sensor that uses TiO_2 is based on the variation in electrical conductivity (resistivity) as a function of the oxygen partial pressure in the gas being analyzed.

The main advantages are

▷ the absence of reference system,

▷ the very significant variation of the conductivity with oxygen partial pressure

$$\sigma \approx 10^{-4}\,S\,cm^{-1} \text{ in an oxidizing environment}$$

$$\sigma \approx 10\,S\,cm^{-1} \text{ in a reducing environment}$$

▷ the possibility of miniaturization,

▷ small fabrication costs.

The main inconveniences are

▷ the properties of conduction evolve with age of the sensor and

▷ the non-specificity of the response.

Solution 5.11 – Amperometric oxygen sensor

1. The operation of an amperometric oxygen sensor is based on determining the diffusion limited current due to the limitation in the oxygen supply at the cathode of a solid-oxide-electrolyte electrochemical cell. The limitation is enforced by a diffusion barrier. The measured limiting current manifests itself in the form of a plateau in the current-voltage curve of the cell. It is related to the oxygen partial pressure in the gas being analyzed.

   The form of the current-voltage curve in the case of a gaseous O_2-Ar mixture is shown in figure 109 and reveals the limiting current I_ℓ.

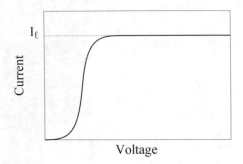

Figure 109 – Form of current-voltage curve of an amperometric oxygen sensor in the case of a gaseous O_2-Ar mixture.

2. Figure 110 shows a schematic of an amperometric sensor in which the oxygen supply is limited by a small-diameter hole.

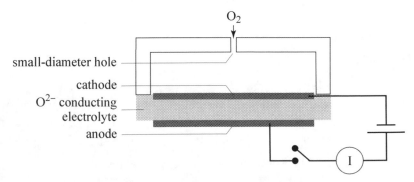

Figure 110 – Schematic of an amperometric sensor in which the oxygen supply is limited by a small-diameter hole.

3. a. In the case of normal diffusion, the relation between the limiting current I_ℓ and the molar fraction x_{O_2} of oxygen in the gaseous mixture is

$$I_\ell = K \times \ln(1 - x_{O_2})$$

K is a constant that depends on the oxygen diffusion coefficient in the gas, the temperature, and the diameter and length of the orifice. Note that the limiting current does not depend on the total pressure of the gas.

b. For diffusion in the Knudsen regime, the relation becomes

$$I_\ell = K' \times P_t \times x_{O_2}$$

where P_t is the total pressure of the gas and K' is a constant.

4. a. ▷ Figure 111 shows the curve for the diffusion-limited current I_ℓ as a function of the oxygen molar fraction at 1 bar. We see that I_ℓ is proportional to $\ln(1-x_{O_2})$.

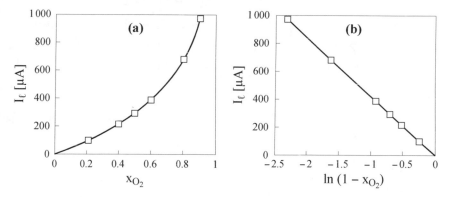

Figure 111 – Diffusion-limited current under 1 bar total pressure as a function of oxygen molar fraction **(a)** x_{O_2} and **(b)** $\ln(1-x_{O_2})$.

The equation for the limiting current is

$$I_\ell = -423 \ln(1 - x_{O_2}) \ [\mu A]$$

▷ Figure 112 shows the curve for the limiting current I_ℓ as a function of oxygen molar fraction at 1.33×10^{-3} bar and 400 °C. We see that I_ℓ is proportional to x_{O_2}.

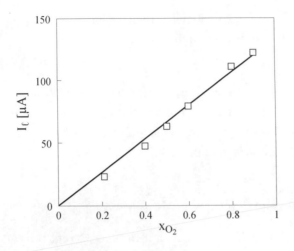

Figure 112 – Diffusion-limited current as a function of oxygen molar fraction x_{O_2} for a total pressure of 1.33×10^{-3} bar.

The equation for the diffusion-limited current is

$$I_\ell = 133.4 \times x_{O_2} \ [\mu A]$$

b. The preceding results show that we are dealing with a normal-diffusion regime when the total pressure and temperature is 1 bar at 400 °C and with a regime of Knudsen diffusion when the total pressure is 1.33×10^{-3} bar at 400 °C.

5. a. Figure 113 shows the curve for the limiting current as a function of temperature.

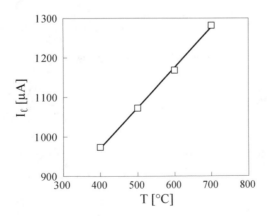

Figure 113 – Limiting current as a function of temperature for an oxygen molar fraction of 0.9 at 1 bar.

b. The equation for the diffusion-limited current is

$$I_\ell = 1.02\,T + 563.5 \ [\mu A]$$

The temperature T is expressed in °C.

6. Figure 114 shows the general form of the response of an amperometric sensor in contact with a gaseous mixture containing oxygen and water vapor. The first step is associated with the oxygen limiting current and the second with that of the water vapor. The current at which these steps occur is directly related to the molar fraction of these species in the mixture.

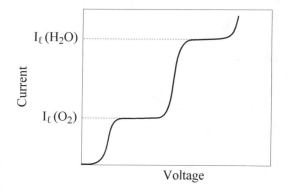

Figure 114 – Form of response of amperometric sensor in contact with a gaseous mixture containing oxygen and water vapor.

7. Drilling a small-diameter orifice in a controlled and reproducible manner proves to be a difficult operation. To avoid this, the system may be improved by depositing a porous layer on the cathode to act as a diffusion barrier against the gas that is being analyzed (fig. 115).

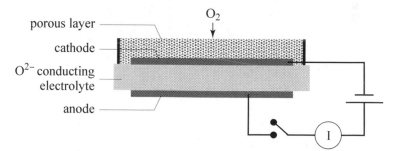

Figure 115 – Schematic of an amperometric sensor in which the oxygen supply is limited by a porous layer.

Solution 5.12 – Coulometric oxygen sensor

1. We begin by extracting by electrochemical reduction all the oxygen from the interior compartment. Under these conditions, we can write $P_{O_2}^{int} \approx 0$. Next we reverse the polarization and we measure the charge required to raise the

pressure $P_{O_2}^{int}$ in the interior compartment to the pressure $P_{O_2}^{ext}$ of the external gas to be analyzed, so that $P_{O_2}^{int} = P_{O_2}^{ext}$. This condition is satisfied when the emf ΔE of the cell goes to zero.

2. The expression relating the oxygen partial pressure of the gas to be analyzed to the charge Q required to equalize the pressure is

$$P_{O_2} = \frac{RT}{4FV} \times Q$$

Numerical evaluation gives

$$P_{O_2} = \frac{8.314 \times 973}{4 \times 96\,480 \times 3.5 \times 10^{-7}} \times 9.68 \times 10^{-3}$$

$$P_{O_2} = 580\,Pa$$

3. The measurement error due to the coulometric sensor is

$$\rho = \frac{0.58 - 0.5}{0.5} \times 100\%$$

$$\rho = 16\%$$

4. Among the advantages, we can cite the following:

▷ no reference electrode and

▷ miniaturization is possible.

The main drawbacks are

▷ the interior volume of the compartment must be determined,

▷ a risk of leaks,

▷ oxygen permeation, and

▷ discontinuous measurement.

Solution 5.13 – Nitrogen oxide sensor

A – Type (I) cell involving sodium ionic conducting electrolyte (NASICON or Na-β-alumina)

1. **a.** Expression of emf for cell (Ia)

$$Na / NASICON (Na^+) / NaNO_3 (porous layer) / Au,NO_2,O_2$$

as a function of NO_2 partial pressure

Consider the following equilibria:

$$Na \rightleftharpoons Na^+ + e$$

$$\mu_{Na} = \tilde{\mu}_{Na^+}^{NASICON} + \mu_e^{Na} - F\varphi_1^{Na}$$

At the NASICON (Na^+) / $NaNO_3$ interface, we have

$$Na_{NASICON}^+ \rightleftharpoons Na_{NaNO_3}^+$$

$$\tilde{\mu}_{Na^+}^{NASICON} = \tilde{\mu}_{Na^+}^{NaNO_3}$$

At the $NaNO_3$ (porous layer) / Au,NO_2,O_2 interface, we have

$$Na^+ + NO_2 + \tfrac{1}{2}O_2 + e \rightleftharpoons NaNO_3$$

$$\mu_{NaNO_3} = \tilde{\mu}_{Na^+}^{NaNO_3} + \mu_{NO_2} + \frac{1}{2}\mu_{O_2} + \mu_e^{Au} - F\varphi_2^{Au}$$

As a first approximation, assume that the chemical potential of electrons is the same in the sodium and in the gold. We obtain the following expression for the emf ΔE_{Ia} of the cell (Ia):

$$\Delta E_{Ia} = \varphi_2^{Au} - \varphi_1^{Na}$$

$$\Delta E_{Ia} = \frac{1}{F}\left(\mu_{Na} + \mu_{NO_2} + \frac{1}{2}\mu_{O_2} - \mu_{NaNO_3}\right)$$

The factor in parentheses is the opposite of the free enthalpy of the following reaction:

$$Na + NO_2 + \tfrac{1}{2}O_2 \rightleftharpoons NaNO_3 \qquad (1)$$

which gives
$$\Delta E_{Ia} = -\frac{\Delta_r G_1}{F}$$

$$\Delta E_{Ia} = \frac{1}{F}\left(-\Delta_r G_1^\circ + RT \ln P_{NO_2} + \frac{RT}{2}\ln P_{O_2}\right)$$

where $\Delta_r G_1^\circ$ is the change in standard free enthalpy of reaction (1) and P_{NO_2} and P_{O_2} are the respective NO_2 and O_2 partial pressures expressed in bars.

b. The expression for emf as a function of the molar fractions x_{NO_2} and x_{O_2}, for the case in which the total pressure is 1 bar, is

$$\Delta E_{Ia} = \frac{1}{F}\left(-\Delta_r G_1^\circ + RT \ln x_{NO_2} + \frac{RT}{2}\ln x_{O_2}\right)$$

2. To verify that the expression for emf is valid, we must compare the theoretical slope s_{th} to the experimental slope s_{expt}.

▷ Theoretical slope $\quad s_{th} = \dfrac{\Delta E}{\Delta \ln x} = \dfrac{RT}{F}$

$$s_{th} = \dfrac{8.314 \times 423}{96\,480} = 36.4 \text{ mV per decade}$$

▷ Experimental slope $\quad s_{expt} = \dfrac{\Delta E}{2.3\,\Delta \log x}$

$$s_{expt} = \dfrac{150 - (-50)}{2.3\,(1.65 + 0.77)} = 35.9 \text{ mV per decade}$$

The slopes s_{th} and s_{expt} are essentially equal, which validates the expression for emf. The number of electrons exchanged is 1.

3. The expression for the change in free entropy of the reaction is deduced from that for the emf

$$\Delta_r G_1^\circ = -F\Delta E + RT \ln x_{NO_2} + \dfrac{RT}{2} \ln x_{O_2}$$

This is evaluated numerically with $x_{NO_2} = 10^{-5}$, $x_{O_2} = 0.21$, and $T = 423$ K. Based on figure 92, for a NO_2 concentration of 10 ppm, we determine a potential difference between the sensor terminals of 91.3×10^{-3} V. We obtain

$$\Delta_r G_1^\circ = -91.3 \times 10^{-3} \times 96\,480 + 8.314 \times 423 \ln 10^{-5} + \dfrac{8.314 \times 423}{2} \ln 0.21$$

$$\Delta_r G_1^\circ = -52.04 \text{ kJ mol}^{-1}$$

4. Expression of emf for cell (Ib)

$$\text{Na / NASICON (Na}^+\text{) / NaNO}_2 \text{ (porous layer) / Au,NO,O}_2$$

as a function of NO partial pressure
We apply the same reasoning as for question 1(a).
Consider the following equilibria:

$$\text{Na} \rightleftharpoons \text{Na}^+ + \text{e}$$

$$\mu_{Na} = \tilde{\mu}_{Na^+}^{NASICON} + \mu_e^{Na} - F\varphi_1'^{Na}$$

At the NASICON (Na^+) / $NaNO_2$ interface, we have

$$\text{Na}^+_{NASICON} \rightleftharpoons \text{Na}^+_{NaNO_2}$$

$$\tilde{\mu}_{Na^+}^{NASICON} = \tilde{\mu}_{Na^+}^{NaNO_2}$$

At the $NaNO_2$ (porous layer) / Au,NO,O$_2$ interface, we have

$$Na^+ + NO + \tfrac{1}{2}O_2 + e \rightleftharpoons NaNO_2$$

$$\mu_{NaNO_2} = \tilde{\mu}_{Na^+}^{NaNO_2} + \mu_{NO} + \tfrac{1}{2}\mu_{O_2} + \mu_e^{Au} - F\varphi_2'^{Au}$$

As a first approximation, assume that the chemical potential of electrons is the same in the sodium and in the gold. This leads to the following expression for the emf ΔE_{Ib} of the cell (Ib): $\Delta E_{Ib} = \varphi_2'^{Au} - \varphi_1'^{Na}$

$$\Delta E_{Ib} = \tfrac{1}{F}\left(\mu_{Na} + \mu_{NO} + \tfrac{1}{2}\mu_{O_2} - \mu_{NaNO_2}\right)$$

The factor in parentheses is the opposite of the free enthalpy of the following reaction:

$$Na + NO + \tfrac{1}{2}O_2 \rightleftharpoons NaNO_2 \qquad (2)$$

which gives $$\Delta E_{Ib} = -\frac{\Delta_r G_2}{F}$$

$$\Delta E_{Ib} = \tfrac{1}{F}\left(-\Delta_r G_2^\circ + RT \ln P_{NO} + \frac{RT}{2}\ln P_{O_2}\right)$$

where P_{NO} and P_{O_2} and the respective NO and O_2 partial pressures expressed in bars.

The expression for emf as a function the molar fractions x_{NO_2} and x_{O_2}, for the case in which the total pressure is 1 bar, is

$$\Delta E_{Ib} = \tfrac{1}{F}\left(-\Delta_r G_2^\circ + RT \ln x_{NO} + \frac{RT}{2}\ln x_{O_2}\right)$$

5. The variations in the emf of cells (Ia) and (Ib) with the corresponding nitrous oxide partial pressures are the same. For (Ib) we deduce

$$s_{th} = 36.7 \text{ mV per decade}$$

The experimental slope is calculated based on figure 92. We obtain

$$s_{expt} = \frac{\Delta E}{2.3\,\Delta \log x}$$

$$s_{expt} = \frac{100 - (-200)}{2.3\,(3.30 + 0.25)} = 36.7 \text{ mV per decade}$$

The slopes s_{th} and s_{expt} are essentially equal, which validates the expression for the emf. The number of electrons exchanged is 1.

6. **a.** Calculate the respective slopes of the curves for the emf of cells (Ia) and (Ib) as a function of O_2 concentration with no NO_2 or NO.

Based on segment (a) of figure 94, we obtain

$$s_{expt} = \frac{\Delta E}{2.3\,\Delta\log x}$$

$$s_{expt} = \frac{100 - (-242)}{2.3\,(3-0)} = 20.6\,\text{mV per decade}$$

for cell (Ia) at 230 °C.

Based on segment (b) of figure 94, we obtain for cell (Ib), at 190 °C,

$$s_{expt} = \frac{-200 - (-300)}{2.3\,(3.200 - 1.097)} = 20.7\,\text{mV per decade}$$

b. The experimental slope is very close to the theoretical slope s_{th} obtained with an electrode reaction involving two electrons; that is,

$$s_{th} = \frac{RT}{2F}$$

▷ for cell (Ia) $s_{th} = \dfrac{8.314 \times 503}{2 \times 96\,480} = 21.6\,\text{mV per decade}$

▷ for cell (Ib) $s_{th} = \dfrac{8.314 \times 463}{2 \times 96\,480} = 19.95\,\text{mV per decade}$

c. For these conditions, we can propose the following electrode reaction:

$$O_2 + 2e \rightleftharpoons O_2^{2-}$$

where O_2^{2-} is the peroxide ion.

7. The proposed reactions should involve a single electron and the peroxide ion O_2^{2-}. We can write

▷ cell (Ia) $NO_2 + O_2^{2-} + e \rightleftharpoons NO_3^- + O^{2-}$

▷ cell (Ib) $NO + O_2^{2-} + e \rightleftharpoons NO_2^- + O^{2-}$

The potential of each of these electrodes is proportional to $\frac{RT}{F}\ln P_{NO_x}$ ($x = 1$ or 2) and does not depend on the oxygen partial pressure.

8. In the presence of only nitrogen dioxide NO_2, a possible reaction involving a single electron and no O_2 is

$$NO_2 + e \rightleftharpoons NO_2^-$$

9. To avoid the drawbacks of a sodium reference electrode, we can use the following electrodes: Ag / AgCl + NaCl / NASICON

$$Ag, O_2 / Na^+\ \beta\text{-alumina}$$

B – Type (II) cell involving solid O^{2-} ionic conducting electrolyte (ZrO_2-Y_2O_3, ZrO_2-MgO)

1. Let us write out the electrochemical reactions

   ▷ at the reference electrode $\quad \frac{1}{2}O_2^{Ref} + 2e \rightleftharpoons O^{2-}$

   ▷ at the measurement electrode $\quad NO^{Mes} + 2O^{2-} \rightleftharpoons NO_3^- + 3e$

2. The emf of the cell is $\qquad \Delta E = E^+ - E^-$

   with $E^+ = E^\circ_{O_2/O^{2-}} + \dfrac{RT}{2F} \ln \dfrac{P_{O_2}^{\frac{1}{2}}}{a_{O^{2-}}}$ and $E^- = E^\circ_{NO_3^-/NO} + \dfrac{RT}{3F} \ln \dfrac{a_{NO_3^-}}{P_{NO}}$

$$\Delta E = E^\circ_{O_2/O^{2-}} - E^\circ_{NO_3^-/NO} - \left(\dfrac{RT}{2F} \ln a_{O^{2-}} + \dfrac{RT}{3F} \ln a_{NO_3^-} \right)$$
$$+ \dfrac{RT}{4F} \ln P_{O_2} + \dfrac{RT}{3F} \ln P_{NO}$$

   Assuming that the activities $a_{O^{2-}}$ and $a_{NO_3^-}$ are constant, we obtain

$$\Delta E = A + \dfrac{RT}{4F} \ln P_{O_2} + \dfrac{RT}{3F} \ln P_{NO}$$

   with $A = E^\circ_{O_2/O^{2-}} - E^\circ_{NO_3^-/NO} - \left(\dfrac{RT}{2F} \ln a_{O^{2-}} + \dfrac{RT}{3F} \ln a_{NO_3^-} \right) = $ const.

Solution 5.14 – The sodium-sulfur battery

A – Solid electrolyte of sodium-sulfur (Na-S) battery

1. The chemical formula for sodium β alumina is Na_2O-$11Al_2O_3$ or $Na_2Al_{22}O_{34}$.

2. The dominant conducting species is the cation Na^+. The heightened ionic conductivity in the family of β alumina is due to the anisotropy of the structure. The sodium moves primarily in conducting planes perpendicular to the $\vec{c}$ axis that have low ionic density, separating the AlO_4 spinel blocks (see fig. 96).

3. Figure 97 shows that, for identical stoichiometry, the best conductivity is obtained with sodium. Exerting a hydrostatic pressure on the crystal in the $\vec{c}$-axis direction results in approaching the spinel blocks together and reducing the space near the conducting plane. This gives preference to the conducting species with smaller cationic radii, such as Li^+.

B – Sodium-sulfur battery

1. Examination of the formula $Na_{1.67}Mg_{0.67}Al_{10.33}O_{17}$ shows that we can write it in the form $Na_{1+x}Mg_xAl_{11-x}O_{17}$ with $x = 0.67$.
   Identification with the formula $(Na_2O)_\alpha(MgO)_\beta(Al_2O_3)_\gamma$ leads to

$$\alpha = \frac{1+0.67}{2} = 0.835 \quad \beta = 0.670 \quad \gamma = \frac{11-0.67}{2} = 5.165$$

The formula is $(Na_2O)_{0.835}(MgO)_{0.670}(Al_2O_3)_{5.165}$

In both forms, we verify that the number of oxygen atoms is 17.

2. We start by calculating the total number n_t of moles of β'' alumina to prepare,

$$n_t = \frac{m}{M} = \frac{\rho V}{M}$$

where M is the molar mass of the compound and V is the volume of the plate. We find $M = 605.6 \, g\,mol^{-1}$ and $V = 9.425 \, cm^3$, which gives $n_t = 5.011 \times 10^{-2}$ mol. Based on the formula $(Na_2O)_{0.835}(MgO)_{0.670}(Al_2O_3)_{5.165}$, we calculate for a formula unit the molar fraction and the number of moles of each oxide. The calculation for Na_2O gives

$$x_{Na_2O} = \frac{0.835}{0.835 + 0.670 + 5.165} = 0.1252$$

$$n_{Na_2O} = x_{Na_2O} \times n_t$$

We obtain $n_{Na_2O} = 6.274 \times 10^{-3}$ mol. By proceeding in the same manner, we obtain $n_{MgO} = 5.034 \times 10^{-3}$ mol and $n_{Al_2O_3} = 3.88 \times 10^{-2}$ mol. From this we deduce

$$m_{Na_2O_3} = 2\, n_{Na_2O} \times M_{Na_2O_3}$$

$$m_{Na_2O_3} = 2 \times 6.274 \times 10^{-2} \times 85$$

$$m_{Na_2O_3} = 1.067 \, g$$

An identical calculation gives

$$m_{Mg(NO_3)_2} = 0.747 \, g$$

$$m_{Al(NO_3)_3} = 16.529 \, g$$

Note – To be absolutely rigorous, we should have accounted for shrinkage during sintering.

3. The electrochemical chain describing the Na/S battery is

$$Na_\ell \, / \, Na_{1.67}Mg_{0.67}Al_{10.33}O_{17} \, / \, S$$

4. At the anode, sodium is oxidized

$$Na_\ell \rightleftharpoons Na^+ + e$$

At the cathode, the two following reactions occur:

$$5S + 2e \rightleftharpoons S_5^{2-}$$

then

$$\left(1 - \tfrac{x}{5}\right)S_5^{2-} + \tfrac{2x}{5}e \rightleftharpoons S_{5-x}^{2-}$$

The overall operating reaction

▷ for the first step is

$$5S + 2Na_\ell \rightleftharpoons Na_2S_5$$

▷ and for the second step is

$$\tfrac{2x}{5}Na_\ell + \left(1 - \tfrac{x}{5}\right)Na_2S_5 \rightleftharpoons Na_2S_{5-x}$$

5. The theoretical energy density is expressed by

$$W_{th} = \frac{Q \times \Delta E}{m}$$

Q is the expected electric charge and ΔE is the potential difference at the terminals of the battery. Taking as reference the global reaction involving the first cathodic reaction, we obtain

$$W_{th} = \frac{2F \times \Delta E}{2M_{Na} + 5M_S}$$

Numerical evaluation gives

$$W_{th} = \frac{2 \times 96\,480 \times 2.08}{[(2 \times 23) + (5 \times 32)] \times 10^{-3}} = 1\,948 \, kJ \, kg^{-1}$$

$$W_{th} = 541 \, W \, h \, kg^{-1}$$

6. In evaluating W_p, we must account for the mass of the electrolyte, the sodium and sulfur containers, the heating system, the thermal isolation apparatus, ... Because more mass must be considered, the specific energy density is thus lower.

7. Figure 116 shows a schematic of a simple Na/S cell in a tubular configuration.

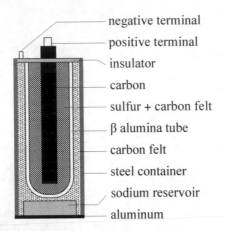

negative terminal
positive terminal
insulator
carbon
sulfur + carbon felt
β alumina tube
carbon felt
steel container
sodium reservoir
aluminum

Figure 116 – Schematic
of a simple Na/S battery.

8. The theoretical capacity of a battery is the maximum charge that it can supply to the exterior. This depends on the amount of active substance present at the electrodes and corresponds to the disappearance of one of the reactants. For a given electrode, the expression for the charge Q is

$$Q = \left| \frac{v_e F}{v_i} n_i \right|$$

where v_e is the stoichiometric coefficient of the electron, v_i is the stoichiometric coefficient of the active species i (Ox or Red), and n_i is the number of moles of the active species.

The electrode with the lowest capacity dictates the capacity of the battery. For the sodium electrode, we have the reaction

$$Na_\ell \rightleftharpoons Na^+ + e$$

or
$$Q_{Na} = \frac{1 \times 96\,480}{1} \times \frac{230}{23} = 964\,800 \text{ C}$$

For the sulfur electrode, the following reaction must be considered:

$$5S + 2\left(1 - \frac{x}{5}\right)e \rightleftharpoons \frac{x}{5}S_5^{2-} + S_{5-x}^{2-}$$

Given that x = 1, we obtain

$$5S + 2.4e \rightleftharpoons 0.2S_5^{2-} + S_4^{2-}$$

The capacity of the sulfur electrode is

$$Q_S = \frac{2.4 \times 964\,800}{5} \times \frac{800}{32} = 1\,157\,760 \text{ C}$$

We observe that the smallest capacity is that of the sodium electrode. Consequently, the capacity of the battery is

$$Q = 964\,800\,C = 268\,A\,h$$

Solution 5.15 – General information on fuel cells

1. A solid-electrolyte fuel cell is an energy-conversion system that involves an electrochemical chain where the oxidizing agent (oxidant) is reduced at the cathode and a fuel (reducer) is oxidized at the anode. The cathode and anode are separated by a purely-ionic-conducting solid electrolyte (polymer, ceramic, glass) that seals and isolates the separate anodic and cathodic compartments.

2. Examples of solid electrolyte fuel cells

   ▷ Example of a solid oxide electrolyte fuel cell
   Currently, the most mature solid oxide electrolyte fuel cell (SOFC) uses the following electrochemical chain:

   $$O_2, LSM / YSZ / Ni\text{-}YSZ, H_2$$

   The combustion agent is the oxygen in air. LSM is the cathode material. It is a solid solution based on oxides with the general formula $La_{1-x}Sr_x\,MnO_{3-\delta}$. YSZ forms the solid electrolyte. This is yttria-stabilized zirconia with the formula $(ZrO_2)_{0.91}(Y_2O_3)_{0.09}$. The current transport is implemented by the oxide ion *via* a vacancy mechanism.
   The anode material consists of a composite cermet, where the ceramic is yttria-stabilized zirconia and the metal is nickel. This material is called cermet Ni-YSZ. It generally consists of 40 vol% of metallic nickel and has a porosity on the order of 30%. H_2 is the fuel. It must be free of sulfur compounds, which poison the anode.
   The fuel cell operates in this configuration from 600 to 1 000 °C.

   ▷ Example of a solid polymer-electrolyte fuel cell
   The proton exchange membrane fuel cell (PEMFC) that currently offers the highest performance uses the following electrochemical chain:

   $$O_2, Pt / \textit{Nafion} / Pt, H_2$$

   The electrode materials are composed of platinum particles deposited on a carbon substrate and that act as catalyst.

The polymer electrolyte *Nafion* is a proton conductor with the formula

$$-[(CF_2-CF_2)_n-(CF_2-\underset{|}{CF})]_x-$$
$$OCF_2-\underset{|}{CF}-CF_3$$
$$OCF_2-CF_2-SO_3H$$

The hydrogen must be free of carbon monoxide, which poisons the platinum. This fuel cell operates from 60 to 90 °C.

▷ Another example is the direct methanol fuel cell (DMFC). It uses the same membrane as electrolyte as the PEMFCs and operates at temperatures ranging from 60 to 80 °C. At the anode, methanol CH_3OH is oxidized at a catalyst layer according to the reaction

$$CH_3OH + H_2O \longrightarrow CO_2 + 6H^+ + 6e$$

and oxygen is reduced at the cathode:

$$O_2 + 4H^+ + 4e \longrightarrow 2H_2O$$

We observe that water is consumed at the anode and produced at the cathode. The overall operating reaction is

$$2CH_3OH + 3O_2 \longrightarrow 2CO_2 + 4H_2O$$

with proton transfer to the membrane. The DMFC does not require *a priori* catalytic reforming of the fuel. However, the diffusion of liquid methanol across the membrane and the need for a water reservoir at the anode are drawbacks, notably in terms of the specific energy of the cell.

3. Molecular hydrogen H_2 does not exist in the free state and must thus be fabricated.

▷ Fabrication of H_2 by reforming methane CH_4 with water vapor

$$CH_4 + H_2O \rightleftharpoons CO + 3H_2$$

in the presence of iron as catalyst, followed by the reaction

$$CO + H_2O \rightleftharpoons CO_2 + H_2$$

catalyzed by Fe_2CO_3, $Cr_2(CO_3)_3$, ...
The balanced reaction equation is

$$CH_4 + 2H_2O \rightleftharpoons CO_2 + 4H_2$$

▷ Fabrication of H_2 by partial oxidation of methane CH_4

Partial oxidation of methane

$$CH_4 + \tfrac{1}{2}O_2 \rightleftharpoons CO + 2H_2$$

in the presence of iron as catalyzer, followed by the reaction

$$CO + H_2O \rightleftharpoons CO_2 + H_2$$

catalyzed by $Fe_2(CO_3)_2$, $Cr_2(CO_3)_3$, ...
The balanced reaction equation is

$$CH_4 + H_2O + \tfrac{1}{2}O_2 \rightleftharpoons CO_2 + 3H_2$$

▷ Electrolysis of water
The overall reaction is

$$H_2O \rightleftharpoons H_2 + \tfrac{1}{2}O_2$$

H_2 and O_2 are produced at the cathode and anode, respectively. The electrolytes may be liquids or solids.

▷ Other procedures
Thermal disassociation of H_2O, photochemical disassociation of H_2O, oxidation of reducing metal, bioproduction by using enzymes, ...

4. a. Figure 117 shows schematically the curve describing the potential difference ΔE and the power density as a function of current density that it delivers.

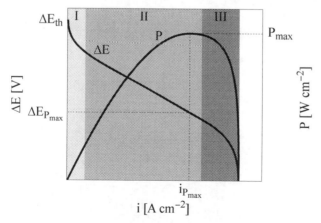

Figure 117 – Form of potential difference ΔE
and power density as a function of the current density.

b. At zero current, we measure the thermodynamic potential difference ΔE_{th}. At low current densities (domain I), ΔE is controlled by the activation overpotentials due to charge-transfer phenomena. At high current densities (domain III), ΔE is controlled by the concentration overpotentials due to diffusion phenomena. In the intermediate (domain II), ΔE is controlled by ohmic voltage-drop phenomena.

The optimal use of the cell corresponds theoretically to the operating point that delivers the maximum power P_{max}. The corresponding current density is thus $i_{P_{max}}$ and is obtained for a given potential difference $\Delta E_{P_{max}}$.

Solution 5.16 – Solid oxide fuel cell (SOFC)

1. Each electrode hosts the following electrochemical half-reaction:

$$\tfrac{1}{2} O_2 + 2e' + V_O^{\cdot\cdot} \rightleftharpoons O_O^{\times}$$

At the negative electrode, we are actually at the equilibrium

$$H_2O + 2e' + V_O^{\cdot\cdot} \rightleftharpoons O_O^{\times} + H_2$$

The electric potentials of the positive and negative electrodes E_c and E_{an} are respectively

$$E_c = E^{\circ}_{O_2/O_O^{\times}} + \frac{RT}{4F} \ln \frac{P^c_{O_2}}{P^{\circ}} \quad \text{and} \quad E_{an} = E^{\circ}_{O_2/O_O^{\times}} + \frac{RT}{4F} \ln \frac{P^{an}_{O_2}}{P^{\circ}}$$

We deduce

$$\Delta E = E_c - E_{an}$$

$$\Delta E = \frac{RT}{4F} \ln \frac{P^c_{O_2}}{P^{an}_{O_2}}$$

with $P^c_{O_2} = 0.21$ bar. We obtain $P^{an}_{O_2}$ based on the formation reaction of water

$$H_2 + \tfrac{1}{2} O_2 \rightleftharpoons H_2O$$

whose equilibrium constant at temperature T is

$$K_T = \frac{P_{H_2O} (P^{\circ})^{1/2}}{P_{H_2} (P^{an}_{O_2})^{1/2}}$$

Using $P^{\circ} = 1$ bar, we have

$$P^{an}_{O_2} = \frac{1}{K_T^2} \left(\frac{P_{H_2O}}{P_{H_2}} \right)^2$$

By writing the relation between the equilibrium constant K_T and the free enthalpy of the formation reaction of water, we have

$$\Delta_f G_T^\circ = -RT \ln K_T = -RT \ln \frac{P_{H_2O}}{P_{H_2}(P_{O_2}^{an})^{1/2}}$$

from which we obtain

$$\ln P_{O_2}^{an} = 2\frac{\Delta_f G_T^\circ}{RT} + 2\ln \frac{P_{H_2O}}{P_{H_2}}$$

The emf is expressed as

$$\Delta E = \frac{RT}{4F} \ln P_{O_2}^c - \frac{RT}{2F} \ln \frac{P_{H_2O}}{P_{H_2}} - \frac{\Delta_f G_T^\circ}{2F}$$

or $\qquad \Delta E = \frac{RT}{4F} \ln P_{O_2}^c - \frac{RT}{2F} \ln \frac{P_{H_2O}}{P_{H_2}} - \frac{\Delta_f H_T^\circ - T\Delta_f S_T^\circ}{2F}$

2. Numerical evaluation at 1 000 K gives

$$\Delta E = \frac{8.314 \times 1000}{4 \times 96\,480} \ln 0.21 - \frac{8.314 \times 1000}{2 \times 96\,480} \ln \frac{0.02}{0.98}$$
$$+ \frac{247\,900 + (1\,000 \times 55.3)}{2 \times 96\,480}$$

$$\Delta E = 1.132 \text{ V}$$

3. The change in emf as a function of temperature depends on the sign of the change in the entropy of the overall operating reaction of the cell, itself given by the change $\Delta n_{(g)}$ of the number of moles of gas. The results are given in table 52.

Table 52 – Change in emf ΔE_T with temperature for cells supplied with various fuels.

Overall reaction	$\Delta n_{(g)}$	Sign of $\Delta_f S$	$\Delta E = f(T)$
$H_{2(g)} + \frac{1}{2}O_{2(g)} \rightleftharpoons H_2O_{(g)}$	-0.5	<0	decreasing
$CH_{4(g)} + 2O_{2(g)} \rightleftharpoons CO_{2(g)} + 2H_2O_{(g)}$	0	0	$\approx$ constant
$CH_3OH_{(g)} + \frac{3}{2}O_{2(g)} \rightleftharpoons CO_{2(g)} + 2H_2O_{(g)}$	0.5	>0	increasing

4. The thermodynamic efficiency of the cell as a function of thermodynamic quantities is expressed as

$$R_{th} = \frac{\Delta_r G_T}{\Delta_r H_T} = \frac{\Delta_r H_T - T\Delta_r S_T}{\Delta_r H_T}$$

$$R_{th} = 1 - T \frac{\Delta_r S_T}{\Delta_r H_T}$$

The oxidation reactions for the various fuels is

$$H_{2(g)} + \tfrac{1}{2} O_{2(g)} \rightleftarrows H_2O_{(g)} \tag{1}$$

$$R_{th1} = \frac{-192.6}{-247.9}$$

$$R_{th1} = 77.7\%$$

$$CH_{4(g)} + 2O_{2(g)} \rightleftarrows CO_{2(g)} + 2H_2O_{(g)} \tag{2}$$

$$R_{th2} = \frac{-800.2}{-800.5}$$

$$R_{th2} = 99.9\%$$

5. The power density P of the cell depends on the emf ΔU under current when the cell carries a load and on the current density i,

$$P = \Delta U \times i = \frac{\Delta U \times I}{S}$$

where S is the active surface area of the electrodes.

When the cell delivers current, the emf diminishes because of the ohmic drop across the electrolyte ($R_{el}I$) and the overpotentials η_a and η_c at the anode and cathode, respectively.

$$\Delta U = \Delta E - |\eta_c| - \eta_a - R_{el}I$$

The electrolyte resistance R_{el} is determined based on the ionic conductivity of YSZ taken from figure 98 (1.3×10^{-2} S cm^{-1} at 1 000 K) and the cell dimensions according to $\quad R_{el} = \dfrac{1}{\sigma} \times \dfrac{\ell}{S}$

$$R_{el} = \frac{1}{1.3 \times 10^{-2}} \times \frac{0.01}{\pi \times 5^5} = 9.8 \times 10^{-3} \, \Omega$$

Consequently, $\quad \Delta U = 1.132 - 0.2 - 0.02 - 9.8 \times 10^{-3} \times 23.5$

$$\Delta U = 0.682 \text{ V}$$

which gives a power density of

$$P = \frac{0.682 \times 23.5}{\pi \times 5^5}$$

$$P = 0.205 \text{ W cm}^{-2}$$

Solution 5.17 – Use of hydrocarbons in SOFCs

1. When operating with hydrogen, we have

   ▷ at the anode $\quad 2H_2 + 2O^{2-} \longrightarrow 2H_2O + 4e$

   ▷ at the cathode $\quad O_2 + 4e \longrightarrow 2O^{2-}$

   The overall reaction is $\quad O_2 + 2H_2 \longrightarrow 2H_2O$

2. Chemical oxidation reactions

   a. ▷ Reforming reaction with water vapor
   - For methane

   $$CH_4 + H_2O \rightleftharpoons CO + 3H_2$$

   - For an arbitrary hydrocarbon C_nH_m

   $$C_nH_m + nH_2O \rightleftharpoons nCO + (n + \tfrac{m}{2})H_2$$

   - Applied to butane C_4H_{10}

   $$C_4H_{10} + 4H_2O \rightleftharpoons 4CO + 9H_2$$

   ▷ Reforming reaction with carbon dioxide
   - For methane

   $$CH_4 + CO_2 \rightleftharpoons 2CO + 2H_2$$

   - For an arbitrary hydrocarbon C_nH_m

   $$C_nH_m + nCO_2 \rightleftharpoons 2nCO + \tfrac{m}{2}H_2$$

   - Applied to butane C_4H_{10}

   $$C_4H_{10} + 4CO_2 \rightleftharpoons 8CO + 5H_2$$

   ▷ Cracking reaction
   - For methane

   $$CH_4 \rightleftharpoons C + 2H_2$$

   - For an arbitrary hydrocarbon C_nH_m

   $$C_nH_m \rightleftharpoons nC + \tfrac{m}{2}H_2$$

   - Applied to butane C_4H_{10}

   $$C_4H_{10} \rightleftharpoons 4C + 5H_2$$

▷ Chemical partial oxidation reaction
- For methane

$$CH_4 + \tfrac{1}{2}O_2 \rightleftharpoons CO + 2H_2$$

- For an arbitrary hydrocarbon C_nH_m

$$C_nH_m + \tfrac{n}{2}O_2 \rightleftharpoons nCO + \tfrac{m}{2}H_2$$

- Applied to butane C_4H_{10}

$$C_4H_{10} + 2O_2 \rightleftharpoons 4CO + 5H_2$$

b. Because of the formation of carbon monoxide and of solid carbon, the additional reactions possible at the anode are

▷ electrochemical oxidation of carbon monoxide

$$2CO + 2O^{2-} \rightleftharpoons 2CO_2 + 4e$$

▷ chemical oxidation of carbon monoxide

$$2CO + O_2 \rightleftharpoons 2CO_2$$

▷ the disproportionation of carbon monoxide (Boudouard reaction)

$$2CO \rightleftharpoons CO_2 + C$$

▷ the water-gas shift reaction

$$CO + H_2O \rightleftharpoons CO_2 + H_2$$

▷ electrochemical oxidation of carbon

$$C + O^{2-} \rightleftharpoons CO + 2e$$

$$C + 2O^{2-} \rightleftharpoons CO_2 + 4e$$

▷ carbon steam reforming

$$C + H_2O \rightleftharpoons CO + H_2$$

3. Direct electrochemical oxidation reactions
  a. Electrode reactions and overall reaction for SOFC-type cell fueled by methane and operating by total electrochemical oxidation and partial electrochemical oxidation

  ▷ Total electrochemical oxidation
  - at the anode

$$CH_4 + 4O^{2-} \rightleftharpoons CO_2 + 2H_2O + 8e$$

- at the cathode $\quad 2O_2 + 8e \rightleftharpoons 4O^{2-}$

The overall reaction is

$$CH_4 + 2O_2 \rightleftharpoons CO_2 + 2H_2O$$

▷ Partial electrochemical oxidation
- at the anode

$$CH_4 + 3O^{2-} \rightleftharpoons CO + 2H_2O + 6e$$

- at the cathode $\quad \frac{3}{2}O_2 + 6e \rightleftharpoons 3O^{2-}$

The overall reaction is

$$CH_4 + \frac{3}{2}O_2 \rightleftharpoons CO + 2H_2O$$

b. For an arbitrary hydrocarbon C_nH_m
  ▷ Total electrochemical oxidation
- at the anode

$$C_nH_m + (2n + \tfrac{m}{2})O^{2-} \longrightarrow nCO_2 + \tfrac{m}{2}H_2O + (4n + m)e$$

- at the cathode

$$(n + \tfrac{m}{4})O_2 + (4n + m)e \longrightarrow (2n + \tfrac{m}{2})O^{2-}$$

The overall reaction is

$$C_nH_m + (n + \tfrac{m}{4})O_2 \longrightarrow nCO_2 + \tfrac{m}{2}H_2O$$

Applied to propane
- at the anode

$$C_3H_8 + 10\,O^{2-} \longrightarrow 3CO_2 + 4H_2O + 20e$$

- at the cathode

$$5O_2 + 20e \longrightarrow 10O^{2-}$$

The overall reaction is

$$C_3H_8 + 5O_2 \longrightarrow 3CO_2 + 4H_2O$$

Application to but-2-ene
- at the anode

$$C_4H_8 + 12O^{2-} \longrightarrow 4CO_2 + 4H_2O + 24e$$

- at the cathode

$$6O_2 + 24e \longrightarrow 12O^{2-}$$

The overall reaction is

$$C_4H_8 + 6O_2 \longrightarrow 4CO_2 + 4H_2O$$

▷ Partial electrochemical oxidation
- at the anode

$$C_nH_m + (n+\tfrac{m}{2})O^{2-} \longrightarrow nCO + \tfrac{m}{2}H_2O + (2n+m)e$$

- at the cathode

$$(\tfrac{n}{2}+\tfrac{m}{4})O_2 + (2n+m)e \longrightarrow (n+\tfrac{m}{2})O^{2-}$$

The overall reaction is

$$C_nH_m + (\tfrac{n}{2}+\tfrac{m}{4})O_2 \longrightarrow nCO + \tfrac{m}{2}H_2O$$

Applied to propane
- at the anode

$$C_3H_8 + 7O^{2-} \longrightarrow 3CO + 4H_2O + 14e$$

- at the cathode

$$\tfrac{7}{2}O_2 + 14e \longrightarrow 7O^{2-}$$

The overall reaction is

$$C_3H_8 + \tfrac{7}{2}O_2 \longrightarrow 3CO + 4H_2O$$

Applied to but-2-ene
- at the anode

$$C_4H_8 + 8O^{2-} \longrightarrow 4CO + 4H_2O + 16e$$

- at the cathode

$$4O_2 + 16e \longrightarrow 8O^{2-}$$

The overall reaction is

$$C_4H_8 + 4O_2 \longrightarrow 4CO + 4H_2O$$

c. Cell emf ΔE
▷ For a hydrocarbon C_nH_m
- total electrochemical oxidation

$$\Delta E = \Delta E^\circ_{C_nH_m/CO_2} + \frac{RT}{(4n+m)F} \times \ln\frac{P_{C_nH_m} \times P_{O_2}^{n+(m/4)}}{P_{CO_2}^n \times P_{H_2O}^{m/2}}$$

- partial electrochemical oxidation

$$\Delta E = \Delta E^{\circ}_{C_nH_m/CO_2} + \frac{RT}{(2n+m)F} \times \ln\frac{P_{C_nH_m} \times P_{O_2}^{(n/2)+(m/4)}}{P_{CO_2}^n \times P_{H_2O}^{m/2}}$$

▷ For an alkane C_nH_{2n+2} $m = 2n + 2$
The preceding expressions become
- total electrochemical oxidation

$$\Delta E = \Delta E^{\circ}_{C_nH_{2n+2}/CO_2} + \frac{RT}{(6n+2)F} \times \ln\frac{P_{C_nH_{2n+2}} \times P_{O_2}^{(3n+1)/2}}{P_{CO_2}^n \times P_{H_2O}^{n+1}}$$

- partial electrochemical oxidation

$$\Delta E = \Delta E^{\circ}_{C_nH_{2n+2}/CO_2} + \frac{RT}{(4n+2)F} \times \ln\frac{P_{C_nH_{2n+2}} \times P_{O_2}^{(3n/2)+1}}{P_{CO_2}^n \times P_{H_2O}^{n+1}}$$

Solution 5.18 – Thermodynamic study of methane reforming in SOFC

1. The reactions to consider to calculate the open-circuit voltage of the cell are the electrochemical reactions

   ▷ at the anode $CH_4 + 4O^{2-} \rightleftharpoons CO_2 + 2H_2O + 8e$
   ▷ at the cathode $\quad 2O_2 + 8e \rightleftharpoons 4O^{2-}$

   The balance equation corresponds to the total combustion reaction of methane and is
   $$CH_4 + 2O_2 \rightleftharpoons CO_2 + 2H_2O$$

   The electromotive force ΔE that appears at the terminals of the cell is due to the electric potential difference φ_e between the two electrodes
   $$\Delta E = \varphi_e^c - \varphi_e^a$$

   where φ_e^c and φ_e^a are the electrostatic potentials of the electrons at the positive and negative electrodes, respectively. Moreover, thermodynamic equilibrium leads to the following equalities:
   $$2\mu_{O_2}^c + 8\tilde{\mu}_e^c = 4\tilde{\mu}_{O^{2-}}^c$$

   and $\quad \mu_{CH_4}^a + 4\tilde{\mu}_{O^{2-}}^a = \mu_{CO_2}^a + 2\mu_{H_2O}^a + 8\tilde{\mu}_e^a$

   for the cathode and anode, respectively.
   By combining these relations and given that
   $$\tilde{\mu}_e = \mu_e - F\varphi_e$$

where F denotes the Faraday constant, we obtain

$$\Delta E = \frac{1}{8F}\left(\mu_{CO_2}^a + 2\mu_{H_2O}^a - \mu_{CH_4}^a - 2\mu_{O_2}^c\right)$$

By expanding the chemical potentials, the preceding equation becomes

$$\Delta E = \frac{1}{8F}\left(\mu_{CO_2}^\circ + 2\mu_{H_2O}^\circ - \mu_{CH_4}^\circ - 2\mu_{O_2}^\circ\right) + \frac{RT}{8F} \times \ln\frac{P_{CH_4} \times \left(P_{O_2}^{(c)}\right)^2}{P_{CO_2} \times P_{H_2O}^2}$$

By using

$$\Delta_c G_{CH_4}^\circ = \mu_{CO_2}^\circ + 2\mu_{H_2O}^\circ - \mu_{CH_4}^\circ$$

and

$$\frac{\Delta_c G_{CH_4}^\circ}{8F} = - \Delta E_{CH_4/CO_2}^\circ$$

we obtain

$$\Delta E = \Delta E_{CH_4/CO_2}^\circ + \frac{RT}{4F} \times \ln\frac{P_{CH_4} \times P_{O_2}^{2(c)}}{P_{CO_2} \times P_{H_2O}^2}$$

where $\Delta_c G_{CH_4}^\circ$ and $\Delta E_{CH_4/CO_2}^\circ$ are respectively the standard free enthalpy of combustion for methane and the standard emf associated to the pair CH_4/CO_2.

2. The standard enthalpies of formation $\Delta_f H_i^\circ$ allow us to calculate the standard reaction enthalpy $\Delta_r H_{298\,K}^\circ$ at 298 K

$$\Delta_r H_{298\,K}^\circ = \sum_i \upsilon_i \Delta_f H_i^\circ$$

The Kirchhoff equation allows us to calculate it at 1 073 K

$$\Delta_r H_{1073\,K}^\circ = \Delta_r H_{298\,K}^\circ + \int_{298}^{1073} (\Delta Cp_i)dT$$

Numerical evaluation gives

$$\Delta_r H_{298\,K}^\circ = - 802.2 \text{ kJ mol}^{-1}$$

$$\Delta_r H_{1073\,K}^\circ = - 775.2 \text{ kJ mol}^{-1}$$

In the same way, based on the absolute standard entropies, we obtain the change in the standard reaction entropy at 298 K

$$\Delta_r S_{298\,K}^\circ = \sum_i \upsilon_i S_{i,298\,K}^\circ$$

We next calculate the absolute standard entropies at 1 073 K

$$\Delta_r S_{1073\,K}^\circ = \Delta_r S_{298\,K}^\circ + \int_{298}^{1073} (\Delta Cp_i)\frac{dT}{T}$$

Numerical evaluation gives

$$\Delta_r S_{298\,K}^\circ = 213.6 + (2 \times 188.7) - (2 \times 205) - 186.2$$

$$\Delta_r S^\circ_{298\,K} = -5.2\,J\,K^{-1}\,mol^{-1}$$

and

$$\Delta_r S^\circ_{1073\,K} = 39.37\,J\,K^{-1}\,mol^{-1}$$

The standard free enthalpy of the reaction is given by

$$\Delta_r G^\circ = \Delta_r H^\circ - T\Delta_r S^\circ = -nFE$$

Consequently,

▷ at 298 K $\Delta_r G^\circ_{298\,K} = -802.2 \times 10^{-3} - 298(-5.2)$

$$\Delta_r G^\circ_{298\,K} = -800.65\,kJ\,mol^{-1}$$

$$\Delta E^\circ_{CH_4/CO_2,\,298\,K} = \frac{800.65 \times 10^{-3}}{8 \times 96\,480}$$

from which $\Delta E^\circ_{298\,K} = 1.037\,V$

▷ at 1 073 K $\Delta_r G^\circ_{1073\,K} = -775.2 \times 10^3 - 1073 \times 39.37$

$$\Delta_r G^\circ_{1073\,K} = -817.44\,kJ \times mol^{-1}$$

from which $\Delta E^\circ_{1073\,K} = 1.059\,V$

Note that temperature only slightly affects the standard emf of the pair CH_4/CO_2.

3. Given the proportion of carrier gas, the steam/carbon (S/C) ratios allow us to calculate the partial pressures of the various gases. The expressions established in the preceding questions allow us to calculate the theoretical emf ΔE_{th} of the cell. The experimental emf ΔE_{expt} are obtained from the polarization curves for i = 0 (see fig. 99). Table 53 gives the results. Note the good agreement between the theoretical values and the experimental results (pressures are in bars).

Table 53 – Partial pressure of gas and theoretical and measured emf of a SOFC fueled by methane for various operating modes.

Operating mode	S/C	P_{CH_4}	P_{H_2O}	P_{CO_2}	P_{O_2}	ΔE_{th} [V]	ΔE_{expt} [V]
Direct internal reforming	1	0.05	0.05	4.10^{-3}	0.21	1.184	1.180
Gradual internal reforming	½	0.066	0.033	4.10^{-3}	0.21	1.209	1.210
	¼	0.08	0.02	4.10^{-3}	0.21	1.237	1.240

4. As shown by the equation

$$\Delta E = \Delta E° + \frac{RT}{4F} \times \ln \frac{P_{CH_4} \times P_{O_2}^{2(c)}}{P_{CO_2} \times P_{H_2O}^2}$$

when P_{H_2O} increases, ΔE decreases.

Furthermore,

▷ water vapor enables the steam reforming reaction, which is rapid and thermodynamically favorable. This reaction leads to the formation of hydrogen and carbon monoxide, which in turn may be electrochemically oxidized. This results in a release of 8 electrons, as is the case for total electrochemical oxidation.

▷ water vapor strongly limits the risk of carbon deposition by cracking or by the Boudouard reaction, which poisons the electrocatalytic sites of the anode. Nevertheless, the strong endothermicity of the steam reforming reaction leads to a thermal gradient at the cell input, which manifests itself as a delamination of the anode due to the significant mechanical stress exerted on the ceramic assembly.

5. The principle of gradual internal reforming is summarized in figure 118. Methane is introduced into the anodic compartment in an over-stochiometric ratio with respect to the water vapor, which initiates the steam reforming reaction. The hydrogen produced is electrochemically oxidized at the anode. This reaction leads to the formation of water vapor which, swept along by the gas flux, will gradually participate in the reforming reaction throughout the anodic compartment.

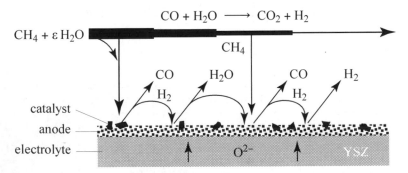

Figure 118 – Principle and schematic of operation of SOFC with gradual internal methane reforming.

In these conditions, the advantages of gradual internal reforming are

▷ a strong decrease in water requirements, which simplifies the system and its use,

▷ delocalization of the thermal gradient due to the endothermicity of the steam reforming reaction,

▷ superior performance due to an increased emf of the cell.

In contrast, the operation without steam at the anode makes the system move vulnerable to carbon deposits.

Solution 5.19 – Electrochemical integrator

1. Electrode (1), which is negatively polarized, is the site of a reduction

$$Ag^+ + e \longrightarrow Ag$$

Electrode (2), which is positively polarized, is the site of an oxidation

$$Cl^- \longrightarrow \tfrac{1}{2}Cl_2 + e$$

2. **a.** The emf at the terminals of the cell is expressed as

$$\Delta E = \frac{RT}{2F} \times \ln \frac{P_{Cl_2}^{(2)}}{P_{Cl_2}^{(1)}}$$

where $P_{Cl_2}^{(1)}$ and $P_{Cl_2}^{(2)}$ denote respectively the chlorine partial pressure in compartments (1) and (2).

$P_{Cl_2}^{(1)}$ corresponds to the equilibrium pressure of chlorine relative to the formation reaction for $AgCl_{(s)}$ $\qquad Ag_{(s)} + \tfrac{1}{2}Cl_{2(g)} \rightleftarrows AgCl_{2(s)}$

where we verify the expression $\qquad P_{Cl_2}^{(1)} = e^{\frac{2\Delta_f G_T^\circ}{RT}}$

$\Delta_f G_T^\circ$ is the standard free enthalpy of the formation reaction of $AgCl_{(s)}$ at temperature T.

$P_{Cl_2}^{(2)}$ is obtained by using the ideal gas law

$$P_{Cl_2}^{(2)} = n_{Cl_2} \frac{RT}{V}$$

where V is the volume of the bulb and n_{Cl_2} is the number of moles of chlorine produced by the current. Denoting by Q the charge that goes through the cell, we can write

$$n_{Cl_2} = \frac{Q}{2F}$$

We thus deduce $\Delta E = \frac{RT}{2F} \ln\left(\frac{Q}{2F} \times \frac{RT}{V}\right) - \frac{\Delta_f G_T^\circ}{F}$

b. Numerical evaluation

The charge Q that goes through the cell is

$$Q = I \times t$$

$$Q = 10^{-2} \times 5 \times 60$$

$$Q = 3\,C$$

and $\Delta E = \dfrac{109.8 \times 10^3}{96\,480} + \dfrac{8.314 \times 298}{2 \times 96\,480} \ln\left(\dfrac{3}{2 \times 96\,480} \times \dfrac{8.314 \times 298}{2 \times 96\,480}\right)$

The emf is $\Delta E = 1.274\ V$

c. For this integrator to operate continuously simply by reading the emf, it requires
   ▷ constant temperature. This requires a thermal management to counter any Joule heating,
   ▷ that the thermodynamic equilibria be in place.

d. The maximum charge Q_{max} that should pass through the cell is obtained from the relations established in 2(a)

$$P_{Cl_2}^{(2)} = n_{Cl_2} \frac{RT}{V} \quad \text{and} \quad n_{Cl_2} = \frac{Q}{2F}$$

or $Q_{max} = \dfrac{2\,FVP_{Cl_2}^{(2)}}{RT}$

$$Q_{max} = \dfrac{2 \times 96\,480 \times 10^{-6} \times 2 \times 10^5}{8.314 \times 298}$$

$$Q_{max} = 15.58\,C$$

Fick's laws of diffusion

Fick's law describes the fluxes and concentrations of a species as a function of time t and of distance x with respect to an origin, which is generally fixed at the electrode-electrolyte interface.

First law

The first law expresses the flux $J_i(x,t)$ due to diffusion of species i at a distance x from an electrode plane

$$J_i(x,t) = -D_i \frac{\partial [i]_{(x,t)}}{\partial x}$$

D_i is the diffusion coefficient of species i.

Second law

The second law expresses that the variation of concentration as a function of time is proportional to the second derivative of concentration as a function of distance x.

$$\frac{\partial [i]_{(x,t)}}{\partial t} = -D_i \frac{\partial^2 [i]_{(x,t)}}{\partial x^2}$$

$[i]_{(x,t)}$ is the concentration of species i.

© Springer Nature Switzerland AG 2020
A. Hammou and S. Georges, *Solid-State Electrochemistry*,
https://doi.org/10.1007/978-3-030-39659-6

Bibliography

S. Baliteau (2005) *Étude et réalisation d'un capteur potentiométrique à dioxyde de carbone en configuration ouverte*, PhD thesis, Université Joseph Fourier, Grenoble, France.

E. Caillot, M. J. Duclot, J. L. Souquet, M. Levy, F. G. K. Baucke and R. D. Werner (1994) A Unified Model for Ionic Transport in Alkali Disilicates below and above the Phase Transition, *Phys. Chem. Glasses* **35**(1):22–27.

C. Déportes, M. Duclot, P. Fabry, J. Fouletier, A. Hammou, M. Kleitz, E. Siebert and J. L. Souquet (1994) *Électrochimie des Solides*, Collection Grenoble Sciences, PUG, Grenoble, France.

M. Duclot (1977) *Méthodes dilométriques appliquées à l'étude du transport électrique dans les cristaux ioniques*, PhD thesis, Université Scientifique et Médicale and Institut National Polytechnique de Grenoble, Grenoble, France.

P. Fabry and C. Gondran (2008) *Capteurs électrochimiques*, Ellipses, Paris.

J. Fouletier, P. Fabry and M. Kleitz (1976) Electrochemical Semipermeability and the Electrode Microsystem in Solid Oxide Electrolyte Cells, *J. Electrochem. Soc.* **123**(2):204–213.

G. S. Fulcher (1925) Analysis of Recent Measurements of the Viscosity of Glasses, *J. Am. Ceram. Soc.* **8**(6):339–355.

S. Georges (2003) *Étude de nouveaux conducteurs par ions O_2 dérivés de $La_2Mo_2O_9$*, PhD thesis, Université du Maine, Le Mans, France.

S. Georges, G. Parrour, M. Hénault and J. Fouletier (2006) Gradual Internal Reforming of Methane: A Demonstration, *Solid State Ionics* **177**(19–25):2109–2112.

A. Hammou (1975) Étude de la conductivité électrique dans le système ThO_2-$YO_{1,5}$, *J. Chim. Phys.* **72**(4):431-438.

P. J. Gellings and H. J. M. Bouwmeester (eds) (1997) *The CRC Handbook of Solid-State Electrochemistry*, CRC Press, Boca Raton, New York, London, Tokyo.

C. Julien and G.-A. Nazri (1994) *Solid State Batteries: Materials Design and Optimization*, Kluwer Academic Publishers, Dordrecht.

© Springer Nature Switzerland AG 2020
A. Hammou and S. Georges, *Solid-State Electrochemistry*,
https://doi.org/10.1007/978-3-030-39659-6

J. H. Kennedy (1977) The β-Aluminas. In: S. Geller (ed) *Solid Electrolytes*, Topics in Applied Physics Vol. 21, Springer-Verlag, Berlin, Heidelberg, New York, p. 105–142.

D. L. Kirk and P. L. Pratt (1967) Ionic Conductivity in Pure and Manganese-Doped Crystals of Sodium Chloride, *Proc. Br. Ceram. Soc.* **9**:215–232.

J.-M. Klein, M. Hénault, P. Gélin ,Y. Bultel and S. Georges (2008) A Solid Oxide Fuel Cell Operating in Gradual Internal Reforming Conditions under Pure Dry Methane, *Electrochem. Solid St. Letters* **11**(8):B144–B147.

J.-M. Klein, M. Hénault, C. Roux, Y. Bultel and S. Georges (2009) Direct Methane Solid Oxide Fuel Cell Working by Gradual Internal Steam Reforming: Analysis of Operation, *J. Power Sources* **193**(1):331–337.

J. A. Labrincha, J. R. Frade and F. M. B. Marques (1993) Defect Structure of $SrZrO_3$, *Solid State Ionics* **61**(1–3):71–75.

V. Levitskii, A. Hammou, M. Duclot and C. Déportes (1976) Influence de l'oxygène sur la conductivité électrique du fluorure de calcium monocristallin, *J. Chim. Phys.* **73**(3):305–314.

S. Lübke and H.-D. Wiemhöfer (1999) Electronic Conductivity of Gd-Doped Ceria with Additional Pr-Doping, *Solid State Ionics* **117**(3–4): 229–243.

N. L. Peterson and C. L. Wiley (1984) Diffusion and Point Defects in Cu_2O, *J. Phys. Chem. Solids* **45**(3):281-294.

M. C. Pope and N. Birks (1977) The Electrical Conductivity of NiO at 1 000 °C, *Corrosion Sci.* **17**(9):747-752.

A. Ringuedé and J. Guindet (1997) Ideal Behavior of a Thin Layer of $La_{0.7}Sr_{0.3}CoO_{3-x}$, *Ionics* **3**(3–4):256-260.

J. Schenin-King (1994) *Processus impliqués dans la chloruration d'oxydes métalliques en milieux chlorures fondus*, PhD thesis, Université Pierre-et-Marie-Curie, Paris, France.

T. Takahashi (1972) Solid Electrolyte Fuel Cells (Theoritics and Experiments). In: J. Hladik (ed) *Physics of electrolytes*, Vol. 2, Academic Press, London, p. 989–1049.

G. Tammann, W. Hesse and Z. Anorg. (1926) Die Abhängigkeit der Viscosität von der Temperatur bie unterkühlten Flüssigkeiten, *Z. Anorg. Allg. Chem.* **156**(1):245–257.

A. Touati and A. Hammou (2006) Determination of the Exchange Current in the SOFC Composite Cathode, *Ionics* **12**(6):339–341.

B. TREMILLON and G. PICARD (1983). In: *Proceedings of the First International Symposium on Molten Salt Chemistry and Technology*, 20–22 April 1983, Kyoto, Japan, p. 93–98.

C. TUBANDT (1932) Leitfahigkeit und Uberfuhrungszahlen in Festen Elektrolyten. In: W. Wiena and F. Harms (eds) *Handbuch der Experimentalphysik*, vol. 12, part. 1, Akademische Verlagsanstalt, Leipzig, Germany, p. 383.

H. VOGEL (1921) Das Temperature-abhängigketsgesetz der Viskosität von Flüssigkeiten, *Phys. Zeit.* **22**:645–646.

M. S.WHITTINGHAM (1976) The Role of Ternary Phases in Cathode Reactions, *J. Electrochem. Soc.* **123**(3):315–320.

M. S. WHITTINGHAM (1973) Experimental Determination of Fast Ion transport in Solids. In: W. Van Gool (ed) *Fast Ion Transport in Solids: Solid State Batteries and Devices,* North Holland Publishing Co., Amsterdam, Netherlands, p. 429–438.

W. ZIPPRICH and H.-D. WIEMHÖFER (2000) Measurement of Ionic Conductivity in Mixed Conducting Compounds using Solid Electrolyte Microcontacts, *Solid State Ionics* **135**(1–4):699-707.

Glossary

Activation energy – The energy required for a chemical reaction (rate constant) or a physical process (conduction, diffusion, …) to occur. It is generally obtained from an Arrhenius plot of the relevant chemical or physical quantity.

Admittance – The admittance Y of an electric circuit is the inverse of the impedance Z of the circuit and is expressed in siemens (S). It is a complex quantity whose real part is the conductance G of the circuit and whose imaginary part is the susceptance B.

Amalgam – In chemistry, an amalgam denotes a metallic alloy. The most common amalgams involve liquid mercury.

Amperometric sensor – An electrochemical device for measuring the concentration of a chemical species by measuring the diffusion limit current (of oxidation or reduction). The sensor must be operated under a current flow.

Anionic Frenkel disorder (pair) – A pair of structure defects formed by thermal agitation that only affect the anionic sublattice of a pure ionic crystal. This is the case for the pair $F_i' + V_F^{\bullet}$ in barium fluoride BaF_2.

Arrhenius coordinates – A representation that consists of plotting the logarithm of a physical quantity as a function of the inverse absolute temperature. It provides information, in particular, about the activated nature of the quantity.

Bode representation – A manner to represent complex impedance spectra that consists of showing, in logarithmic coordinates, the change in an impedance component (modulus, argument, real part, or imaginary part) of an electric circuit or of an arbitrary sample as a function of angular frequency. In contrast to the Nyquist representation, this representation is more often used by physicists.

Brouwer diagram – The representation, in logarithmic coordinates, of the concentration of structure defects and of electronic species (electrons and holes) in a pure or doped M_aX_b compound as a function of the partial pressure of species M or of species X_2. It has a strong analogy with the acid-base and the complexation equilibria present in solution.

Catalyst – A chemical compound that increases the rate of a chemical reaction without modifying the mass balance of the reaction. It participates in several

© Springer Nature Switzerland AG 2020
A. Hammou and S. Georges, *Solid-State Electrochemistry*,
https://doi.org/10.1007/978-3-030-39659-6

elementary steps in the reaction and is regenerated at the end of the reaction. In electrochemistry, it is in contact with the electrode material and its action leads to an increase in current across the electrode. It can be susceptible to poisoning by chemical species present in reaction medium. For example, carbon monoxide CO constitutes a poison for platinum, which is a catalyst for the oxidation reaction of hydrogen.

Cationic Frenkel disorder (pair) – A pair of structure defects formed by thermal agitation that only affect the cationic sublattice of a pure ionic crystal. An example is the pair $Ag_i^{\bullet}$ and V'_{Ag} in silver iodide AgI.

Cermet – A compound formed from a ceramic matrix and a metal.

Complex (cluster) – A complex is the association of two or more ionic defects of different effective charge. The defects involved in complex formation are often mobile species whose "trapping" diminishes the concentration and, consequently, the ionic conductivity of the material. Complex formation is an activated phenomenon observed at low temperature.

Complex impedance spectroscopy – An electric or electrochemical characterization technique used notably in solid-state electrochemistry. It measures the response of an electrochemical system to an ac voltage perturbation of varying frequency and small amplitude around a stationary operating point. It allows, in particular, the study of electrical and dielectric properties of materials and the separation of the various electrical and electrochemical contributions to the behavior of complex systems.

Composite – A solid material made of several non-miscible phases. Mechanically, it consists of a reinforcement and a matrix. The mechanical properties are mainly due to the reinforcement whereas the matrix serves to link and protect the reinforcement with respect to the environment, the cohesion of the composite, and the transmission of forces. The composite thus constituted possesses properties not offered by the individual constituents. In solid-state electrochemistry, a composite is made of a blend of electrolyte and electrode materials. It allows for a gradual transition between the properties of the constituents (type of conduction, dilatation coefficient).

Constant phase element (CPE) – A passive dipole that can behave as a pure resistance, capacitance, or inductance, or any intermediate combination. A CPE is used to model the electric properties of ionic materials that exhibit a strong dispersion in relaxation times, which does not allow them to be modeled as pure elements.

Coordination number – The number of nearest neighbors of an atom in a solid.

Coulometric sensor – Electrochemical device for measuring the concentration of a chemical species by measuring the current.

Counter electrode, reference electrode, working electrode – A basic cell for electrochemical measurements (see figure below) that involves an electrolyte and three electrodes:

▷ the working electrode WE whose properties we want to study by exploiting its characteristic potential E_{WE}/current I,

▷ reference electrode (or comparison electrode) RE whose potential is fixed at E_{RE}, and

▷ the counter electrode CE that closes the circuit, thereby allowing current to flow through the cell.

The potential E_W of the working electrode is obtained based on the measured potential difference $\Delta E = E_{WE} - E_{RE}$. One must not forget to correct for the ohmic drop; in other words, to subtract from ΔE the ohmic drop $R_{el}I$ due to the electrolyte between the working electrode and the reference electrode.

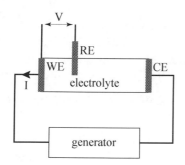

Schematic of the basic cell for electrochemical measurements.

Delamination – The partial or total loss of adherence of a deposit to its substrate.

Decentering angle – In the Nyquist representation, the plots that represent the variation of the opposite of the imaginary part of the complex impedance as a function of the real part take the form of semicircles. Each semicircle is associated with a circular arc whose center is under the abscissa. The decentering angle is the angle of the inflection of the semicircle associated with the arc measured with respect to the real axis.

Defect ionization – The reaction of a structure element with an electronic species. For example, the species $V_O^{\bullet}$ results from the reaction of an oxygen vacancy $V_O^{\bullet\bullet}$ with an electron e'.

Dilatocoulometry – A technique to measure the cationic transport number in solid electrolytes with simple formulas. It is based on the displacement of an electrode-electrolyte interface due to the disappearance of matter by electrolysis.

Disproportionation (equilibrium) – The reaction equilibrium involving three species of the same chemical element with different degrees of oxidation. For example, for iron, $3Fe^{2+} \rightleftharpoons 2Fe^{3+} + Fe$. The forward and reverse directions indicate respectively the disproportionation reaction of Fe^{2+} and the comproportionation (or anti-disproportionation) reaction of Fe^{2+}.

Doping – An operation consisting of introducing in an intentional and controlled manner a quantity of a foreign constituent (dopant) into a host (matrix). This operation can considerably modify certain physical properties of the host. The doping level is very low in materials for microelectronics ($<10^{-6}$). It reaches the at% level in solid-state electrochemistry.

Effective charge – A charge attributed to a structure element (normal or defect) in an ionic crystal. It depends on the real charge of the species that occupies a given site and on the charge of the species that normally occupies the site in a pure ionic crystal. The effective charge of a structure element can be positive, denoted "˙," negative, denoted "′," or zero, denoted "×." The electric neutrality of an ionic crystal involves the effective charges of the species present in the crystal.

Electrical conductivity – The electrical conductivity of a material is due to the displacement of electric charges (electrons, ions) when the material is subjected to a gradient in the chemical potential or in the electrical potential. Its units are siemens/meter [$S\,m^{-1}$]. We generally use siemens/centimeter [$S\,cm^{-1}$].

Electrical mobility – The speed of displacement of a charged species per unit electric force.

Electrical resistivity – An intrinsic property of a material that represents its capacity to oppose the passage of an electric current. Its value is determined by measuring the resistance R of a portion of the material with well-defined geometry and is expressed in ohm meters ($\Omega\,m$).

Electrochemical chain – A system consisting of two electrodes separated by one or more electrolytes. An example is $Ag/AgI/C, I_2$.

Electrochemical coloration – A change in color observed upon injecting electronic charge carriers into a highly polarized electrolyte. The color change is attributed to the formation of color centers, which absorb in the visible light.

Electrochemical integrator – Device to measure a time delay after an electrochemical response.

Electrochemical mobility – The speed of displacement of a charged species per unit electric field.

Electrochemical pump – A device to extract or enrich a gas in a gaseous mixture in a container of freely circulating. Schematically, it consists of a solid electrolyte that isolates the gas to be enriched or purified. The solid electrolyte is used as an ionic conductor with the charge carriers being the ions of the chemical species to be introduced into or extracted from the gas. For example, stabilized zirconia, which is a oxide ion (O^{2-}) conductor, is used to electrochemically pump oxygen.

Electrode – A system consisting of an electronic conductor (metal, semiconductor, electrode material) in contact with an electrolyte (aqueous, organic, molten salt, ceramic, polymer) and that is the site an oxidation-reduction reaction. An electrode can be at equilibrium or polarized. In the latter case, it carries a current characterized by the sign in electrochemistry. If the current is positive, the electrode hosts an oxidation reaction and constitutes the anode. If the current is negative, it is the cathode and hosts a reduction reaction.

Electrode overpotential (see Polarization) – The potential difference between a polarized electrode (i.e., one carrying a current) and its equilibrium potential. Contrary to the polarization, an overpotential concerns a well-defined electrochemical reaction. Its variation with current gives information about the type of reaction occurring at the electrode (transfer, adsorption, diffusion, …).

Electrode polarization – Polarizing an electrode consists of imposing a potential E_i different from the equilibrium potential $E_{i=0}$. When polarized, the electrode carries a current of the same sign as the polarization. Thus, a cathode always carries a negative current whereas an anode always carries a positive current, by convention. Often the notation Π is used to denote the potential difference $\Pi = E_i - E_{i=0}$. The polarization, unlike the overpotential, is defined independently of the nature and number of reactions at the electrode.

Electrolyte – An electrolyte is a condensed phase (liquid or solid) in which the electric conductivity is exclusively due to ions (cations, anions). In a solid electrolyte, the electric current is generally due to a single ionic species. For example, the current in stabilized zirconia (YSZ) consists of migrating oxide ions O^{2-}.

Electroneutrality equation – Expresses the electric neutrality of a phase. In solid-state electrochemistry, one must consider the effective charges and the charge of the electronic species.

Electronic conductivity – This is the conductivity due to the electronic species (electrons and holes) present in a material. We speak respectively of n-type or p-type conductivity when the mobile species is the electron or the hole. In metals, the electronic conductivity is a decreasing function of temperature. In semiconductors (electrode materials), electronic conductivity is an activated phenomenon.

Equilibrium exchange current (density of) – Consider an electrode that hosts an equilibrium reaction

$$Ox + ne \rightleftarrows Red$$

In reality, the electrode hosts two opposing reactions of unequal rates. To each reaction is associated an exchange-current density whose absolute value i_0 is the density of the equilibrium exchange current.

Equivalent electrical circuit – The equivalent electrical circuit is a mathematical model obtained by combining pure dipolar electrical elements (R, C, L) and is used to determine the properties of a sample by mathematically adjusting its parameters at experimental test points. This adjustment can be done, for example, by a least-squares fit.

Extrinsic defect – A structure defect due to the presence of impurities or dopants in an ionic crystal. Its definition requires a notation for the introduction of the impurity or dopant in the pure crystal.

Formula unit – The simplest and most succinct formula to denote a chemical compound. For example, the formula unit associated with titanium dioxide is TiO_2 instead of Ti_2O_4 or $Ti_{16}O_{32}$, which are also valid in terms of stoichiometry.

Free volume model – The free volume model is a microscopic model developed to explain the mechanical or electrical behavior of amorphous electrolytes or polymers. From an electric viewpoint, the mobility of charge carriers is said to be "activated" below the ideal glass transition temperature T_G and "assisted" by entropic redistribution above that temperature. The theoretical models used to explain this behavior are the activated hopping model below T_G and the Vogel-Tammann-Fulcher (VTF) model above T_G.

Fuel cell – A system for converting chemical energy into electrical energy that involves the following chains: oxidant, electrode / electrolyte / electrode, reducer. Oxidants such as oxygen or chlorine constitute oxidizing agents, and reducers such as hydrogen, methane, or methanol constitute fuels.

Geometric factor – For a sample of simple geometry (e.g., cylinder, parallelepiped, …), it is the ratio between sample thickness and the area over which is deposited the electrodes. It connects the electrical resistance, which is measurable, to the conductivity of a phase.

GIR – Abbreviation meaning "gradual internal reforming."

Glass transition temperature – The glass transition temperature T_G corresponds to the change from a liquid state (supercooled) to a solid state (metastable) and occurs when the viscosity of the liquid attains the conventionally agreed upon value of 10^{12} Pa s. The temperature at which this transition occurs cannot be precisely defined because it depends on the speed with which the liquid is cooled (or quenched). We thus speak of the glass transition interval.

Impedance – The response of an electric circuit or of a material to an alternating electric current. The impedance of a circuit or a material is defined as a complex number whose modulus is the ratio of the amplitude of the applied voltage to the resulting current, and whose argument is given by the phase difference between the current and the voltage.

Intensiostatic (galvanostatic) mode – A technique for plotting the characteristic potential/current of an electrode. In intensiostatic mode, a constant current is imposed on the working electrode and we measure the potential with respect to a reference electrode.

Ionic conductivity – The conductivity due to ionic species (cations, anions) present in a material. In an aqueous medium, an organic medium, and molten salts, cations and anions contribute to the electric current. In solid-state electrolytes, the conductivity is primarily due to a single ionic species (cationic or anionic). In the solid phase, we associate the displacement of ions to a given mechanism (vacancy, interstitial, ...). For example, in yttria-stabilized zirconia (YSZ), the ionic conductivity is due to oxide ions O^{2-} following a vacancy mechanism. In sodium chloride NaCl, the species Na^+ is responsible for the ionic transport following a vacancy mechanism. Finally, note that ionic conductivity is an activated phenomenon.

Insertion material – An electrode material into which, through a chemical or electrochemical reaction, ions can be inserted. This property results from the particular crystalline structure of the material, which contains empty sites that can host an insertion species. Insertion materials are generally solid solutions of transition-metal oxides. They must have a mixed conductivity with a non-negligible ionic contribution to the conductivity. The insertion reaction is a delocalized reaction that occurs throughout the bulk of the material. Over the last decades, lithium- and proton-insertion materials have received the most attention for research and development. Two examples are $LiMn_2O_4$ and V_2O_5.

Interstitial position – An atom is said to occupy an interstitial position when it occupies a site that is normally empty in the ideal structure.

Intrinsic defect – A structure defect that is due to thermal agitation in a pure ionic crystal. Intrinsic defects often come in pairs. For example, in silver chloride AgCl, intrinsic defects are silver vacancies and interstitial silver cations Ag^+.

Kröger-Vink notation – A notation used in solid-state chemistry and solid-state electrochemistry. It is used to identify the chemical species and filling anomalies in a crystal (vacancies, interstitial ions, ...). This notation clearly identifies, for a given ion, the site it occupies, the chemical nature of the ion, and its effective charge. For example, $O_F^{\bullet}$ refers to an oxide ion with a positive effective charge occupying the site of a fluorine ion.

Limit current – Consider an electrode that hosts the following reduction reaction:
$$Ox + ne \longrightarrow Red$$
If we impose on the electrode a sufficiently negative potential with respect to the equilibrium potential to cancel the concentration of the Ox species at the electrode-electrolyte interface, the current density approaches an asymptote called the limit current. Its value depends on the diffusion coefficient toward the interface of the Ox species. The existence of a limiting current is manifested by a horizontal line in the current-potential curve.

Lattice parameters – In crystallography, the lattice parameters define the dimensions of the unit cell of the crystal. A triclinic system, which is the least symmetric, requires six parameters (a, b, c, α, β, γ) to define the lattice, whereas only a single parameter is needed to define a cubic system. In a cubic system, the lattice parameter denotes the size of the cube associated with the unit cell.

Mass energy density – The mass energy density of a battery corresponds to the quantity of energy that the battery can deliver per unit mass. It is expressed in $Wh\,kg^{-1}$ if the energy is furnished strictly in the form of electricity.

Migration energy – The energy required for an electric field to displace a species from one site to a neighboring site in a crystal.

Mixed conductivity – In solid-state electrochemistry, for given temperature and environmental conditions, a material has mixed conductivity when none of the contributions of ions and of other electronic species (electrons, holes) may be neglected.

Nyquist representation – A way of representing the magnitude of a complex electrical quantity (impedance, admittance, …) as a function of frequency. This representation shows the opposite of the imaginary part as a function of the real part of the given quantity. For impedance, the resulting diagram contains semi-circles, loops, and lines that reveal the behavior of the elements within the electrical circuit under study. The Nyquist representation is commonly used by electrical engineers and electrochemists.

Perovskite – Originally, a perovskite denoted a mineral formed with calcium titanate. Today, it denotes a family of solids with a structure of the same name with the general formula ABO_3 and commonly found in nature. It can have various physical properties (giant magnetoresistance, superconductivity, mixed conductivity, …).

Polarization resistance of an electrode – The polarization resistance of an electrode is the sum of the resistance of each elementary process contributing to the overall reaction at the polarized electrode (adsorption, desorption, charge transfer, diffusion).

Potentiometric sensor – Electrochemical device for measuring the activity of a chemical species by measuring the potential difference across an open circuit. The corresponding elementary cell involves one or more electrolytes inserted between two electrodes, which often serve as current collectors.

Potentiostatic mode – A technique for plotting the characteristic potential/current of an electrode. In potentiostatic mode, the potential difference between the working electrode and the reference electrode is fixed and the current through the working electrode is recorded.

Power density – For an electrochemical generator, the power density is the maximum power that the system can supply at the active surface of the electrodes. It is given in $W\,cm^{-2}$.

Relaxation frequency – A characteristic frequency observed during polarization of dipoles by an alternating electric field when the dielectric loss factor is maximal. For a given phenomenon, it depends only on temperature. The inverse of the relaxation frequency is the relaxation time. The relaxation frequency and relaxation time are strongly correlated with the electrical conductivity.

Schottky disorder (pair) – A pair consisting of a cationic vacancy and an anionic vacancy formed by thermal agitation in a pure ionic crystal. An example is the pair $V'_{Na} + V^{\bullet}_{Cl}$ sodium chloride NaCl.

Sitoneutrality – Expresses the conservation of the ratio of the number of anionic (cationic) sites to the number of cationic (anionic) sites in an ionic crystal. In a crystal with the formula M_aX_b, the sitoneutrality expression is

$$\frac{\text{Number of sites M}}{\text{Number of sites X}} = \frac{a}{b}$$

Solid solution – In thermodynamics, a solid solution is a mixture of pure components forming a homogeneous solid. These components are often of a mineral phase (metals, ionic crystals). The properties of a solid solution can differ greatly from those of the constituent bodies. For example, this is the case of doped silicon-based semiconductors in microelectronics.

Stabilized zirconia – A solid solution of oxides involving zirconium dioxide ZrO_2 and an oxide dopant such as MO or M_2O_3. The presence of the oxide dopant stabilizes the fluorine-type cubic structure over a large range of temperature, composition, and oxygen partial pressure. The most often used material is yttria-stabilized zirconia with the formula $(ZrO_2)_{0.92}$-$(Y_2O_3)_{0.08}$, for its ionic-conduction properties.

Steam reforming – Steam reforming is a catalytic chemical reaction that, with the help of steam, converts hydrocarbons to dihydrogen and carbon monoxide.

For methane, steam reforming occurs in two stages:

▷ a first conversion by a highly endothermic reaction

$$CH_4 + H_2O \rightleftharpoons CO + 3H_2$$

▷ a second slightly exothermic step called the "water gas shift"

$$CO + H_2O \rightleftharpoons CO_2 + H_2$$

These reactions occur at high temperature (800 °C).

Stoichiometry (departure from) – Consider an ionic crystal with the formula M_aX_b resulting from the equilibrium

$$aM + \frac{b}{2}X_2 \rightleftharpoons M_aX_b$$

In reality, the composition of this crystal varies with temperature and its surroundings and the formula may be written in the following different forms: $M_aX_{b-\beta}$, $M_aX_{b+\beta}$, $M_{a+\alpha}X_b$, $M_{a-\alpha}X_b$ where α and β represent the departure from stoichiometry. In $M_aX_{b-\beta}$ and $M_{a-\alpha}X_b$, the crystal is deficient in X and M, respectively. In $M_aX_{b+\beta}$ and $M_{a+\alpha}X_b$, the crystal has excess X and M, respectively. The departure from stoichiometry is manifested by the presence of point defects (vacancies, interstitial ions) whose role is primordial in electrical conductivity, ionic diffusion, …

Structure defect – A site filled with an anomaly with respect to the perfect crystal. In the Kröger and Vink notation, it is defined by its site, the chemical species that occupies the site, and its effective charge.

Structure element – In an ionic crystal, a structure element denotes a chemical entity defined by its site, the element that occupies the site, and the effective charge of the entity. It is said to be normal if it corresponds to the perfect crystal and is otherwise referred to as a defect. For example, in the solid solution CaF_2-KF where KF is the dopant, potassium substitutes for calcium. The species $Ca_{Ca}^{\times}$ and $F_F^{\times}$ are normal structure elements whereas the species K_{Ca}' and $V_F^{\bullet}$ are structure defects.

Substitution position – An atom is said to occupy a substitution position in a solid when it occupies a site normally occupied in the ideal structure by an atom of a different chemical nature.

Tafel equation – Expresses the overpotential η of an electrode as a function of the current density in the electrode as follows: $\eta = a + b \log |i|$
a and b are constant that give access, in particular, to the exchange-current density i_0. This law is verified for slow electronic transfer reactions, when the electrode is subjected to a strong polarization.

Theoretical capacity of a battery – The theoretical capacity of a battery is the electric charge that a fully charged battery can deliver in a single discharge cycle. It is expressed in A h.

Thermodynamic efficiency – The thermodynamic efficiency of a system characterizes the ratio between the real efficiency (i.e., measured) and the maximum theoretical efficiency calculated from thermodynamic quantities. The thermodynamic efficiency of any real system is less than unity because of the irreversible and entropic character of all real transformations.

Transport number – The transport number of charge species i gives the contribution of this species to the total electrical conductivity of a given phase.

Unit cell – The smallest volume of a crystal that preserves the geometric, physical, and chemical properties of the crystal. It gives in particular information on the spatial arrangement of atoms in the crystal. The entire crystal consists of periodic reproductions of the unit cell.

Vacancy – Structure defect denoting an empty crystallographic site in an ionic crystal. It is generally represented by the letter V.

Warburg diffusion impedance – An electrochemical impedance that describes the behavior of an electrode whose current is limited by diffusion phenomena.

Index

© Springer Nature Switzerland AG 2020
A. Hammou and S. Georges, *Solid-State Electrochemistry*,
https://doi.org/10.1007/978-3-030-39659-6

Printed in the United States
By Bookmasters